# GERMAN FIREFIGHTING VEHICLES IN WORLD WAR II

*The Supreme Commander of the Fire Police: Reichsführer SS Heinrich Himmler.* **Photo: LVF Niedersachsen.**

# GERMAN FIREFIGHTING VEHICLES IN WORLD WAR II

Michael Foedrowitz

**Schiffer Military History**
Atglen, PA

Translated from the German by Dr. Edward Force.
This book was originally published under the title,
*Feuerwehrfahrzeuge im Einsatz 1939-1945*,
by Podzun-Pallas Verlag.


Library of Congress Catalog Number: 96-69999.



Printed in China.
ISBN: 0-7643-0191-8



Published by Schiffer Publishing Ltd.
77 Lower Valley Road
Atglen, PA 19310
Phone: (610) 593-1777
FAX: (610) 593-2002
Please write for a free catalog.
This book may be purchased from the publisher.
Please include $2.95 postage.
Try your bookstore first.

# CONTENTS

# FOREWORD

This book describes the service of the German firefighters and their vehicles during World War II. It is not the author's intention to provide a complete list of all firefighting vehicles that saw service and thereby to add another to the already numerous technical publications in existence. As a historian, the author would like to place the emphasis on historical events, in the occupied areas in the East and West, on the fronts, and in Germany itself.

The organization of firefighting units in the Ordnungspolizei under RFSS Himmler brought forth an alarming closeness to the unjust system of the National Socialists, and the problematic side of this was already made glaringly obvious in the synagogue burnings of 1938. The organization and militarization of the firefighters also meant that they saw service in the war: against partisans in the East, in the Warsaw uprising in the summer of 1944, on the Oder and Rhine in 1945, as well as in the final combat around Berlin and many other cities toward the end of the war. As last-minute participants in these combat actions, often equipped only with captured guns and a handful of bullets, considerably more firefighters fell in action than in all the air raids on all the cities of the German Reich.

Almost never before had the German civilian population had so much to thank any civil officers for as the firefighters of World War II. The highest praise is deserved by the firefighters, who worked almost every day from 1942 on, often during air attacks, to their physical and psychological limits, risking their lives to rescue civilians from burning houses and bunkers and rubble-filled cellars. Psychological burdens in view of the difficult work, the many victims and their helplessness in the face of the unrestrained powers of nature in the firestorms placed tremendous pressure on them.

The efficiency of firefighting and lifesaving services was improved considerably by the pooling of all firefighting vehicles and equipment; this guaranteed their quick arrival at the scenes and access to the fires.

It may be hoped that the firefighters of Germany never again will have to see service in the catastrophe of wartime action.

## Organisation der Feuerwehren (Stand 23.10.1943)

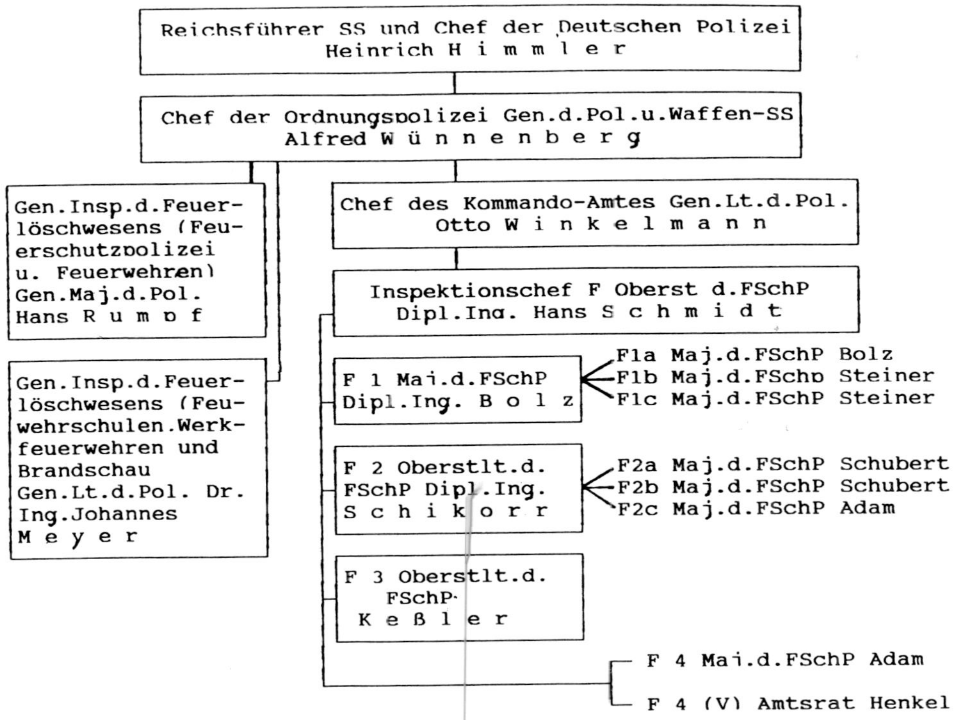


## Fahrzeugbezeichnungen

| RLM | FSchP | laut RdErl.RFSS v.30.4.1943 |
|---|---|---|
| KzS 8 | LLG | LF 8 |
| KS 15 | SLG | LF 15 |
| KS 25 | GLG | LF 25 |
| | LDL | DL 17 |
| | SDL | DL 22 |
| KL 26 | | DL 26 |
| | GDL | DL 32 |
| | SSK | S 3 |
| SKw | GSK | S 4.5 |
| Ts 1.5 | | TFL 15 |
| Ts 2.5 | | TFL 25 |
| | | SRK |
| | | GRK |
| | | LGK |
| Tgsa 8 | | TSA |

# ACKNOWLEDGEMENTS

For their help in the creation of this book, I would like particularly to thank those people without whose advice and factual knowledge a high degree of information would not have been attained. First, though, thanks are due to my wife Isabella, as well as my family, Alice Foedrowitz and Maria Nizio, who kept problems off my back through their understanding, patience and help.

I would particularly like to thank the following people (in alphabetical order), who made important contributions to the success of this book with their interest and factual knowledge: Mr. Klaus Dietz (BF Berlin), Mr. Hagen Ebel (BF Hannover), Mr. Dieter Heimberg (BF Oldenburg), Mr. Hans Hische (FF Celle), Ms. Hoffmann (BAK), Mr. Wolfgang Hornung (Author of a standard work and recognized authority on firefighting matters), Mr. Johanssen (Journalist, Gummersbach), Mr. Rudolf Joop (Leader BF Hannover, retired), Dr. Hermann Koebe, Mr. Metzger (Metz firm archives), Mr. Detlef Nase (Pasewalk), Mr. Mirko Nowak (graphic arts), Mr. Udo Paulitz (well-known author), Mr. Klaus Rabe (Publisher of "Historic Motor Travel"), Mr. Richard Schwarz (Magirus-Iveco firm archives), Dr. Thomas Wolf (Curator, German Firefighting Museum, Fulda), Mr. Wolfgang Zur (BF Berlin and colleague at BF Tegel).

In addition, thanks go to: Mr. Bügel (Henkel KGaA), Ms. Bühring (BLHA), Mr. Dahlström (Luckenwalde), Mr. & Mrs. Gastell (Munich), Mr. Giller (BF Kassell), Mr. Gras (BF Koblenz), Dr. Greffrath (Wolfsburg City Museum), Mr. Hagen (Beuster), Dr. Harrach (Oldenburg), Mr. Albrecht Helmerding (BF Hannover, retired), Ms. Klein (Prussian Cultural Photo Archives, Berlin), Mr. Knigge (FM-Hannover), Mr. Koch (BF Wolfsburg), Mr. Kreter (BF Göttingen), Mr. Nickel (StA Braunschweig), Ms. Oberdieck (Hannover), Mr. Schilling (FF Ahlem), Mr. Schmidt (LFV Sachsen-Anhalt), Mr. Schrape (Oldenburg), Ms. Stekel (Berlin), Mr. Strassl (German Museum, Munich), Mr. Stratmann (LFS Celle), Mr. & Mrs. Tack (Bonn), Mr. Tarrach (FM-Dannenberg-Tramm), Ms. Thomas (StaA Bremen), Mr. Tiesel (StA Bonn), as well as the Daimler-Benz AG (Stuttgart), Adam-Opel AG (Rüsselsheim) and VW-AG (Wolfsburg).

*The LF 25, LF 15 and LF 8 fire trucks with TSA.* Photo: Metz Archives

# The Firefighting Vehicle and Equipment Industry

Before World War II, the German firefighting equipment industry embodied German quality workmanship and outstanding technology in this product area, not only in Germany but throughout the world. The firefighting equipment industry was represented by renowned firms, such as the Carl Metz Firefighting Equipment Factory, founded in 1842, and the company founded by Conrad Dietrich Magirus in 1964 and bearing his name. The leadership of the Metz firm, which was located in Karlsruhe since 1909 and directed by Karl and Alfred Bachert since 1905, developed the first fire engine before World War I. Soon an advanced aerial-ladder technology was developed, and in 1921 the first fully automatic, machine-driven aerial ladder was put on the market. The partner of Metz that produced the chassis was Mercedes-Benz (merged with Daimler in 1926), with its production facility in Gaggenau.

The ladder technology was developed further and brought the firm numerous pioneering German patents, proudly marked on the Metz ladders with the letters D R P (Deutsches Reichspatent). Metz built what was then the longest aerial ladder, with a climbing distance of 44 meters, as well as the first all-metal ladders, which were sold all over the world, including in the Soviet Union and China.

The loss of World War I, the Treaty of Versailles, social and political unrest and economic crises in the Weimar Republic brough almost all German firms to the brink of ruin. The motor vehicle manufacturers had to curtail this lucrative branch of their business, as the city treasuries were drastically short of money for the purchase of firefighting equipment. This meant an increasing competition in a new market situation, with Daimler-Benz aggressively leading the way. In 1928, Magirus came into financial difficulties, with internal decisions plus the neglect of Diesel engine production making a new business concept necessary. The German Bank proposed a merger plan with the Daimler-Benz AG, but it was declined, since the firefighting-equipment branch had not been given sufficient consideration. The corporate goal of Mercedes-Benz, to eliminate Magirus as a competitor with the help of the German Bank, did not succeed. But there was also another side of the matter: Magirus, like their competitors, had produced trucks for the armed forces during World War I, and in 1915 they had been able to hire Mercedes' most talented engi-

# Firefighting Equipment and Vehicle Manufacturing Industry as of the end of 1939.

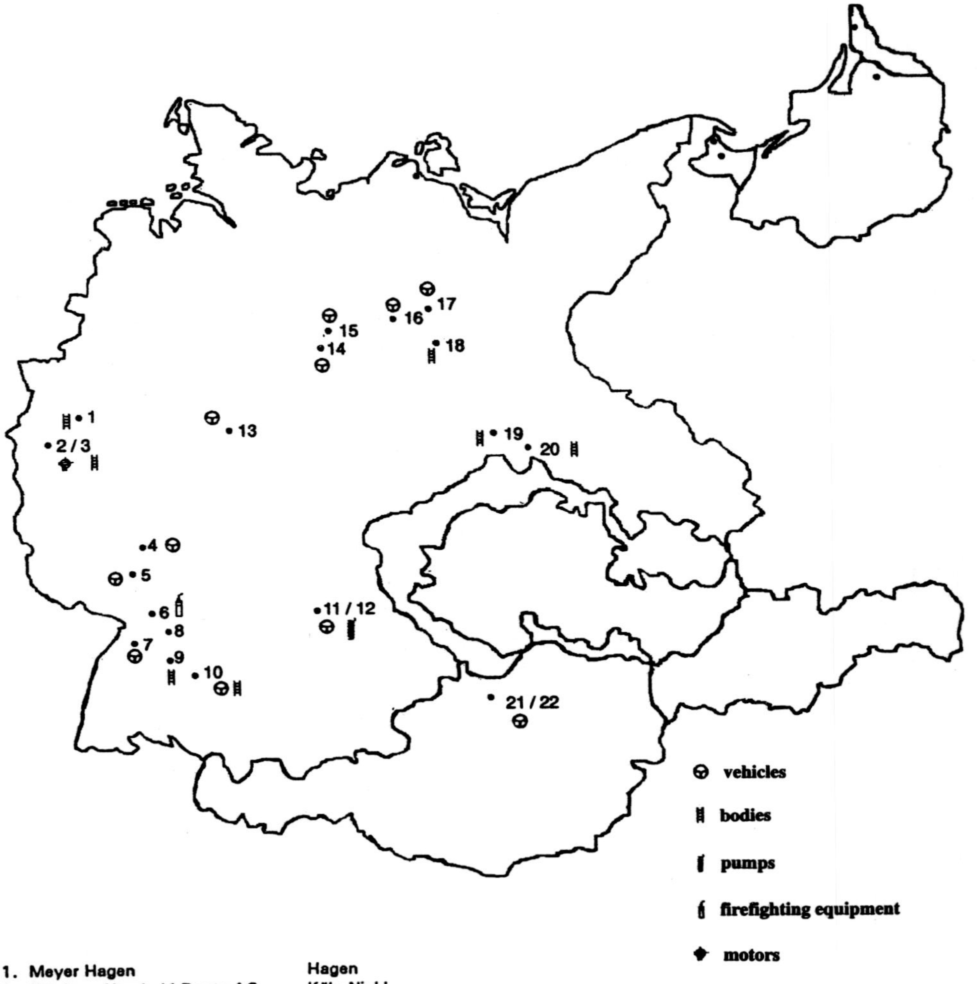


| | |
|---|---|
| 1. Meyer Hagen | Hagen |
| 2. Klöckner-Humbold-Deutz AG | Köln-Niehl |
| 3. August Hoenig | Köln |
| 4. Adam Opel Ag | Rüsselsheim |
| 5. Mercedes Benz | Mannheim |
| 6. Metz | Karlsruhe |
| 7. Mercedes Benz | Gaggenau |
| 8. Bacher | Heilbronn |
| 9. Ziegler | Gingen |
| 10. Magirus | Neu-Ulm |
| 11. MAN | Nürnberg |
| 12. Amag-Hilpert | Nürnberg |
| 13. Henschel & Sohn | Kassel |
| 14. Büssing-NAG AG | Braunschweig |
| 15. Volkswagen | Wolfsburg |
| 16. Adam Opel AG | Brandenburg/Havel |
| 17. Mercedes Benz | Marienfelde |
| 18. Koebe | Luckenwalde |
| 19. Nowak | Bautzen |
| 20. Fischer | Görlitz |
| 21. Konrad Rosenbauer | Linz |
| 22. Steyr-Daimler-Puch | Linz |

*The main gate of Factory I of Magirus in Ulm.* Photo: Iveco-Magirus Archives.

*1941: Bodies of fire trucks on the Magirus assembly line. The wooden frames for the LF 15 are shown here being roughed out.* Photo: Iveco-Magirus Archives.

*The already covered frames for the LF 15 bodies.* Photo: Iveco-Magirus Archives.

*Series production of the LF 15 at Magirus.* Photo: Iveco-Magirus Archives.

*The LF 15 on the assembly line.* Photo: Iveco-Magirus Archives.

*Assembly of S 3000 chassis in Hall 19 at Magirus.* Photo: Iveco-Magirus Archives.

*Mass production of tank-truck chassis. Photo taken January 20, 1943.* Photo: Iveco-Magirus Archives.

*The M 40 chassis (short form) for the fire department in Bergen, Norway.* Photo: Iveco-Magirus Archives.

*The chassis of the KS 25 fire engine based on the Magirus F.S. 145. This vehicle was intended for the Fire Police in Brandenburg. Photo taken November 7, 1938.* Photo: Iveco-Magirus Archives.

*The Magirus military truck chassis.* Photo: Iveco-Magirus Archives.

neer, Heinrich Buschmann, chief designer of the commercial vehicle division in Gaggenau, to put a profitable truck production program into action for Magirus. The leadership of the Daimler-Benz AG soon allied itself with Hitler and the Nazi Party, who promised the "motorization of Germany" as part of their political program.

Since Daimler-Benz could not drive their domestic competitors out of the market, they became part of a "Truck Cartel" in 1930-31 (with, among others, Büssing-NAG and MAN), which might at least limit the involvement of foreign capital and foreign firms. This hit Ford and the Adam-Opel AG (GMC) hardest.

Mercedes secured the Metz firefighting equipment factory for itself: on January 1, 1933 a manufacturing and sales contract between the two companies went into effect. In this contract, Metz agreed not to build any chassis or to offer bodies to other vehicle builders, but to sell only firefighting vehicles on Mercedes chassis. The Daimler-Benz AG in turn agreed to buy all of its firefighting vehicle bodies from Metz and to give up the building of fire pumps.

Meanwhile, the C. D. Magirus AG was fighting for its survival. In 1935, the Klöckner-Humbold-Deutz firm was in search of a new chassis producer for its KHD motors. In that year a contract was signed by Magirus and KHD that included cooperation and joint development of vehicles. On March 6, 1935 KHD took possession of the C. D. Magirus firm, which was first relieved of all financial difficulties.

An unpleasant surprise was prepared for Daimler-Benz by the Adam-Opel AG. On April 7, 1935, probably the most modern truck factory in the world, costing 14 million Reichsmarks, was opened in Brandenburg/Havel, setting new standards for truck production. Here trucks were built in mass production and proved to be superior to the many competing products on account of their reliability (1939: almost 28,000 Opel Blitz vehicles!) With that, Opel became the chief supplier of commercial vehicles for the coming militarization. Until the works were destroyed in an air raid on August 6, 1944, 130,000 three-ton trucks of all types were produced! As early as November 18, 1935, after only 180 days of construction, the first Opel Blitz truck rolled off the assembly line. A daily production run of 150 vehicles was planned. In 1936, the production of three-ton trucks was begun, and the entire truck production was moved from Rüsselsheim to Brandenburg/Havel. The yearly total capacity went up to 25,000 units.

By this time, MAN and KHD had developed a new fuel-injection Diesel engine, putting Daimler-Benz production in this area in question. In 1939 Magirus built its first three-ton truck with all-wheel drive, and Opel vehicles of the same class won more and more of the market.

Meanwhile, the firefighting apparatus industry continued to produce: in 1938 Metz delivered four turntable aerial ladders with a height of 45 meters, these having been ordered directly by Reich Chancellor Adolf Hitler via the Ministry of the Interior. These were the longest turntable aerial ladders with sets of metal ladders, but also the heaviest ladders (Mercedes-Benz LD 6500 chassis with a gross weight of 14.5 tons).

After September 1, 1939 the Air Ministry's ban on producing fire vehicles in Karlsruhe went into effect. At the same time, all materials, half - and finished products were to be transferred to KHD in Ulm. The Baden Economic Ministry, though, had announced heavy fines for shifting. The Army High Command, on the other hand, allowed no contracts to be given to firms near the borders. Before the war began, Metz tried to establish two acceptable locations. By order of the Air Ministry, KHD became the chief contractor for chassis and organized the Magirus-Metz GmbH in Ulm, without obtaining permission from Metz. After long, tedious negotiations it could be agreed that the "fusion" of Magirus and Metz into an exporting company would be postponed during the war. On the basis of this association, Mercedes canceled the contract of 1933. On November 12, 1940, after long negotiations, a new contract, valid until December 31, 1945, was signed.

Daimler-Benz suffered its greatest setback in terms of the production of standardized three-ton trucks. The man in charge, Colonel Adolf von Schell, who was in charge of standardizing measures for truck production, had called on Director Kissel of Mercedes-Benz to become an advisor in the economic group of the motor vehicle industry. In the subsequent "Schell Plan", Daimler-Benz was favored considerably, as Schell left 21 of the 133 (according to other sources, 198) of the trucks placed on the market free. "Although the Daimler models were clearly inferior to the new uniform types, and even more so to the carefully planned Opel three-ton truck with its all-terrain gearbox, its greatly enclosed axles, its ease of conversion to all-wheel drive and, above all, its optimal relation of its own weight to its load limit, Daimler-Benz had to withdraw only its 6.5-ton trucks and limit its two-ton trucks to exporting. The production

*The Koebe KS 25 uniform fire engine on DB L 3750 D chassis. The picture shows the delivery of the trucks and the acceptance commission at factory II.* Photo: Koebe Firefighting Technology, Wenzel publishing, Marburg.

*KzS 8 with Koebe body in front of Koebe Factory II in Luckenwalde.* Photo: Koebe Firefighting Technology, Wenzel Publishing, Marburg.

and further development of 1.5-, 3- and 4.5-ton trucks was still allowed." Schell was particularly generous as concerned three-ton trucks. A new model, with the designation LGF 3000, was put into production in Mannheim at the beginning of 1939, and later also at Untertürkheim. Only after extensive improvements to its design was it accepted by the Army Weapons Office. But the chassis and Diesel engine were still much too heavy, and the engine design was clearly inferior to that of the MAN injection motor in terms of fuel consumption, load limit and starting procedure. The Mercedes-Benz three-ton truck was, as the management itself admitted in September 1940, a failure compared to the competing product made by MAN, and even more so to that made by Opel. Even inside the factory ridicule was heard: the vehicle was "so heavy that it needed all its motive power to get itself moving without a load." The vehicle was still being produced when it became clear, in the Russian war in 1941, that the "LGF 3000 suffered more than its competitors under the rigors of Russia."

The director of the factory in Brandenburg, Heinrich Nordhoff, became the director of the Special 3-ton Truck Branch. For production of the Opel 3-ton truck, Borgward in Bremen (later ruled out) and Mercedes in Mannheim were suggested. The directors of the firm had to grit their teeth and accept; they had not been asked. In August 1942, the contract was signed in Rüsselsheim. DB had to pay a one-time sum of 800,000 Marks plus 120 Marks per truck. This most important license agreement led to troubles in the USA after the war, seriously affecting GMC.

Until the conversion of the Mannheim factory from April to June 1944, after it had been delayed again and again for a variety of reasons, 27,700 LGF trucks were built. In August 1944 the first sixty-five "Opel" trucks were completed.

# The Vehicles

*LLG made by the Koebe firm in Luckenwalde being turned over after the type program for LLG on Feb. 16, 1940. They have 1.5-ton chassis, fuel injection and TSA.* *Photo: Koebe Firefighting Technology, Wenzel Publishing, Marburg.*

*Koebe Fire Engines with front-mounted pumps for the "Reichswerke Hermann Göring AG", July 1938.* *Photo: Koebe Firefighting Technology, Wenzel Publishing, Marburg.*

*Open Koebe body with front pump on 3-ton Opel chassis, for the Neuendorf Aircraft Works.* *Photo: Koebe Firefighting Technology, Wenzel Publishing, Marburg.*

*MB factory fire department with a 1.5-ton MB crew and equipment truck towing Type HZ F 2 fire pump trailer of 1925-28 vintage, with a pumping capacity of 22 HP at 2000 liters per minute. The eight-man crew sat in the open, unprotected.* Photo: MB Archives.

# THE VEHICLES

The KzS 8, called "Katze" (cat) by the firefighters, was ordered in large numbers from 1936 on by the Air Ministry and the Safety and Rescue Service, but the Fire and Rescue Troops and the Fire Police were also equipped with them (in 1938, for example, 175 units were made by Rosenbauer for the Luftwaffe). The bodies were provided by Magirus, Metz, Meyer-Hagen, Rosenbauer, Koebe and others. The 1-and 1.5-ton Opel chassis were used, as was the Mercedes-Benz L 1500 chassis. The KzS 8 was equipped with a TSA. When production began, they were completely open, and even on February 10, 1941 an open KzS 8 of the IV. Firefighting Platoon drove from Braunschweig to Hannover to be used there after an air raid! Later versions had a canvas cover for the crew area, later a closed cab and a cloth cover. The KzS 8 was the forerunner of the later LF 8. After the end of the war, there were still over 300 vehicles of this type in service.

*August 29, 1939: a KzS 8 of the Air Ministry, with Berlin registry for the Tempelhof area.* Photo: Iveco-Magirus Archives.

*1941: Assembly-line pro duction of LF 8 bodies.* Photo Iveco-Magirus Archives.

*LF 8 trucks on Mercedes-Benz chassis with Koebe bodies and trailers, ready to be delivered.* Photo: Koebe Firefighting Technology, Wenzel Publishing, Marburg.

*Koebe special vehicle for use at airfields, Opel chassis from production at the Brandenburg factory.* Photo: Koebe Firefighting Technology, Wenzel Publishing, Marburg.

*This LF 8 on Mercedes chassis was used by the fire police in Berlin.* Photo: FMB.

*An LF 8 on Opel chassis, photographed April 5, 1944.* Photo: Iveco-Magirus Archives.

# The "Light Firefighting Group Vehicle LLG"

The prerequisite for building the LLG was that a light vehicle with a limited load limit was needed that could carry a firefighting group with its equipment to a place of action. Within the farmework of standardizing, the LLG was to be a 1.5-ton truck with a gross weight, including the firefighting group, of 3650 kilograms. A 60-horsepower engine was planned. Since the dimensions were too small to include room for a motor pump, a portable pump was mounted on a trailer. The particulars for the building of the LLG were specified in February 1940. By 1941, all the larger firms in the fire apparatus industry were already occupied with the series production of this vehicle. They used the 1.5-ton Mercedes-Benz Type L 1500 F (standard) chassis, which was the only chassis of this class that had been released for fire truck production at the beginning of the war. The 1.5-ton chassis, standardized by the "Schell Plan", was available as of mid-1941. The most striking difference from the previous 1.5-ton chassis was the decreased wheelbase, from the former 3500 mm to 3000 mm. The earlier 45-HP engine (L 1500 chassis until 1943) showed certain weaknesses, especially when used on mountain roads, but this was to be eliminated by the 60-HP engine with the L 1500 S chassis (produced autumn 1941-1944). The normalized fuel consumption was to be limited to 16 liters per 100 kilometers. Afterward came the three-ton Opel Blitz with a 60-HP carbureted engine. There was an all-wheel-drive version of the Opel Blitz. The cost was 12,500 to 14,000 Reichsmarks.

The new standardized LLG's also were planned to include extensive equipment for the driver and group leader of the firefighters.

From February 28 to March 2, 1941 a test drive with standardized vehicles was held in the Salzburg area, on snowy and icy roads, with numerous upgrades up to 30%.

The LLG (LF 8 as of 1943) was built from 1940 to 1943. After 1945, a great number (over 1000 units) of the LF-8 Mercedes-Benz L 1500 S continued to see service with volunteer firefighters until into the seventies. In all, 3850 of the LF 8 were built.

*A KzS 8 with TSA trailer with side hose reels. Note the application of the national emblem.* Photo: FMB.

*Koebe fire engine (half-body type), developed along with the Volunteer Firefighting Office in 1939. Note the partly covered suction hose in the trailer as well as the side hose reels covered with canvas.*
*Photo: Koebe Firefighting Technology, Wenzel Publishing, Marburg.*

# The LF 8 & TSA, LF 15 & LF 25 Fire Trucks

For the LLG and KzS 8, motor pump trailers were produced, since the trucks themselves could not hold the 800 liter/minute pump. Magirus built a "Modell Goliath G 36" trailer, which had a weight of 380 kilograms with pneumatic tires. The two hose reels combined could carry 270 meters of Type C hose. Metz built a trailer without external hose reels, and Koebe brought out a trailer in completely enclosed form, with a canvas cover.

The TSA was not successful in front action in Russia, as the hitch often broke under the rough conditions there.

In the type program, the TSA 1008 design became the standard. Including body and load, it could weigh up to 1000 kilograms. As a result of the rubber shortage, wheels with steel tires and wooden rims could also be provided.

Single-axle trailers were also made in other versions: as a hose trailer (with B and C hose), as an electric trailer with a portable generator (producing 5 kilowatts) for an equipment truck, as an equipment trailer with tools, lifting tackle, etc., as a forest-fire trailer with forest-fighting apparatus, and as the AL 12, AL 17 and AL 22 ladder trailers.

*TSA production by Magirus.* Photo: Iveco-Magirus Archives.

*A Magirus TSA, "Goliath" model, in 1941.* Photo: FMB.

*LF 15 (Metz body) with swiveling water cannon.* Photo: Metz Archives.

*A KS 15 on Henschel chassis, ordered by the RLM, seen on November 11, 1938.* Photo: Iveco-Magirus Archives.

*The first SLG test truck with a pressed-wood body, seen on April 14, 1943.* Photo: Iveco-Magirus Archives.

*An FlKS 15 on 3-ton uniform chassis by Opel for the RLM, seen on April 30, 1941.* Photo: Iveco-Magirus Archives.

*An FlKS 15 on KHD FS 330 chassis for the RLM, seen on July 29, 1942.* Photo: Iveco-Magirus Archives.

*A KS 25 on Mercedes-Benz 3-4.5-ton chassis, Type Ls 3750, with a 6-cylinder OM 67/3 Diesel engine producing 100 HP. The pump of this fire truck had a capacity of 2500 liters per minute at a height of 80 meters. The crew numbered eleven men.* Photo: Daimler-Benz Archives.

The standard large firefighting group vehicle (GLG), later LF 25, was a further development of the KS 25 and used a 4.5-ton chassis of at least 6800 kilograms. Its price was about 35,000 RM. From 1935 to 1942, over 1000 units are believed to have been built, and in all, 2650 LF 25 trucks were built. The improved version of the LF 25/36 was produced as of 1943 (LF 25/43) and supplied in small numbers to LS (mot.) units as well as the Fire Police.

The Mercedes LS 3750 or KHD FS 145 chassis was used for the GLG, as were the Mercedes-Benz L 4500 S and KHD S 4500 less often. The vehicle proved itself well and was still used in Greece and Crete, among others, until the end of the seventies. Some 500 of the LF 25/36 were used by German fire departments after 1945.

*The "long nose" of Magirus: This KS 25 fire engine with Ml45 chassis was used by the Dortmund Fire Police and photographed on January 18, 1939. On the roof are suction hose, ladders, folding stretcher, etc.* Photo: Iveco-Magirus Archives.

*Equipment and tool chests of the KS 25.* Photo: Iveco-Magirus Archives.

*KS 25 on MB L4500S chassis withg Metz body.* Photo: Metz Archives.

*LF 25 (GLG) on Magirus S4500 chassis for the Hamburg Fire Police, photographed December 1, 1942.* Photo: Iveco-Magirus Archives.

*Koebe SS 25 with CO2 extinguishing system, on Büssing chassis.* Photo: Koebe Firefighting Technology, Wenzel Publishing, Marburg.

*A KS 25 of the Berlin Fire Police, with the standard set of equipment.* Photo: FMB.

*A KS 25 of the SHD (#1492), made for the Air Ministry in Berlin.* Photo: Iveco-Magirus Archives.

*A KS 25 on Daimler-Benz L4000 chassis, made for the Air Ministry in Berlin, photographed August 2, 1941.* Photo: Iveco-Magirus Archives.

*Here the long engine hood of the M145 is especially clear to see.* Photo: Iveco-Magirus Archives.

*An LF 25 of 1938 ? with Metz body. It belonged to the Karlsruhe Air-Raid Police (license Pol 60539). Note the 40 km disc on the rear door. The vehicle carries two lengths of suction hose, a folding stretcher, ladders, and a set of cutters that were used for fuel-tank fires.* Photo: Paulitz/Metz Archives.

*A Metz DL 17 on Opel 1.5-ton chassis with pump trailer.* Photo: Metz Archives.

*Metz LDL on Mercedes-Benz L 1500 S chassis, built around 1943.* Photo: Metz Archives.

# Aerial Ladders

Before the war, in the thirties, there were seven aerial ladders of varying heights in general use: the DL 17, 22, 26, 30, 32, 36 and 46. The production figures of the Metz firm alone show that from 1933 to 1939, 289 DL units were built, more than 1/4 of which were sold outside Germany. In 1940, 71 were made, in 1941, only 59. The norm and type program left only three types during the war: the DL 17, 22 and 32. According to Brunswig, 60, 210 and 40 of these types were built during the war. In the course of war production, a ladder branch was founded, led by Dipl. Ing. Alfred Bachert of Metz and his deputy, Dipl. Ing. Oscar Herterich of KHD, as of October 1, 1943. While Metz alone was responsible for the DL 17 and 22, KHD and Metz both built the DL 32.

According to the regulations for the building of fire trucks, the DL 17 was called LDL, and carried a three-piece, all-steel ladder with a climbing height of 17 meters. A 1.5-ton Mercedes-Benz (L 1500 or L 1500 S) chassis was used. The DL 22 was used on the Magirus GFS 145 and S 4500 chassis. The Metz ladders of this version were mounted on Mercedes-Benz LD 3750 chassis. The GDL (DL 32) had a Magirus GFS 145 or S 4500 chassis, and the Metz ladders had the Mercedes-Benz LD 3750 chassis.

The numerous KL 26 Luftwaffe ladder trucks generally used the Mercedes-Benz L 3750 and FL 145 chassis until the middle of 1941.

*A DL 22, photographed on September 10, 1943.* Photo: Iveco-Magirus Archives.

*GDL for the Hamburg Fire Police, photographed March 25, 1943.* Photo: Iveco-Magirus Archives.

*Ladder Truck No. 1 of the Kassel Fire Police during training in 1934 (Metz on MB chassis).* Photo: Kassel BF Archives.

*A DL 36 for sale abroad. The cab was not yet enclosed, but roofed and protected from wind.* Photo: Iveco-Magirus Archives.

*Two hose reels on a Metz DL 26 ladder truck of the Berlin BF.* Photo: FMB.

**From the training regulations for LS aerial ladder units:**

118. **Climbing:** Before climbing the ladder, the truck driver and engineer always have to make sure that the engine is turned off, all jacks are set evenly on their struts, and the ladder is erected properly. The ladder drive remains engaged. Standing free at a 70-75-degree angle, each section of the ladder can hold one man when the ladder is fully extended. With an attached C pressure hose attached, though, the ladder can only be used by two men on separate sections of the ladder.

Standing free at an angle of less than 65 degrees, the ladder can be used by only one or two men until the load indicator begins to sound.

The extended and placed (leaned) ladder can be climbed by 12 to 15 men at the same time, but this is only allowable when the ladder was set up "soft."

Several men must not climb the ladder in step, but at irregular rhythms, to avoid making the ladder sway.

If the man standing on the rungs of the ladder is handed a tool, the upper man should always climb down to the next lower man, and then climb back up after receiving the object.

An occupied ladder nust not be moved (turned, tilted, raised, extended or retracted) or driven.

*A KL 26 Metz ladder on MB Type LD 3750 chassis.* Photo: Metz Archives.

*A DL 22, built in 1943, on Daimler-Benz L3000F chassis.* Photo: Paulitz/Metz Archives.

*A DL 26, built for the Luftwaffe in 1939, with Metz ladder. Note the bumper and front third of the fender, already painted white.* Photo: Metz Archives.

*Metz ladder trucks with front-end pumps at a Dresden Fire Department demonstration in 1935. These trucks filled the Dresden Fire Department's order for six aerial ladder trucks from Metz in 1934.* Photo: Metz Archives.

*Seven DL 26 built by Metz on MB chassis at a ladder demonstration before the Karlsruhe Concert Hall.*
Photo: Metz Archives.

*The KL 46 with its ladder raised.* Photo: Metz Archives.

*The DL 46 is climbed by a Fire Police officer in Berlin.* Photo: Dietz Collection.

*DL 46, erected and extended.* Photo: Sig. Dietz.

*A Mercedes-Benz LD6500 chassis (6-cylinder OM 57 Diesel engine with 150 HP at n = 1700 rpm, 12,517 cc displacement). The six-section fully automatic all-steel ladder made by Metz could be extended to 45 meters.* Photo: Metz Archives.

*An LF 25 of Magirus FS145 of the Brandenburg/Havel Fire Police, and a Magirus KL 26.* Photo: Iveco-Magirus Archives.

*A DL 26 + 2 aerial ladder on Magirus M45 LG chassis, of the Aachen Fire Police (8/2/1938) with a Magirus KS 25, Type M45, both with front pumps.* Photo: Iveco-Magirus Archives.

# Tanker-Pumpers and TFL

*A Ts 1.5 tank pumper for the Luftwaffe (photo 1/19/1943) on Opel Blitz test vehicle 3.6-6700 A (all-wheel drive).* Photo: Iveco-Magirus Archives.

*A TFL 15 on uniform chassis (Opel Blitz 3-ton, all-wheel drive), photographed 5/25/1944.* Photo: Iveco-Magirus Archives.

*Ts H 315 tanker-pumper for the "Hermann Goring Works" fire department in Salzgitter-Watenstedt. Note the outside seats! Photographed 6/30/1941.* Photo: Iveco-Magirus Archives.

*TSH 515 (Henschel) tanker-pumper, photgraphed 5/10/1943. This TFL 15/43 was delivered to a fire police regiment.* Photo: Iveco-Magirus Archives.

Magirus built the TFL in great numbers for export (USSR, Rumania, Belgium). In 1939 the cab was enclosed. From 1936 to 1942, the Air Ministry ordered a great number for airport fire departments (Kfz. 343), with various bodies. Henschel delivered large numbers of the Ts. 2.5 or Ts. 2.5a on the off-road 33 FA 1 chassis, with Metz and Magirus bodies. Between 1936 and 1940, 600 Ts. 2.5 (with 2500-liter tanks plus 250 liters of foam) were bought by the Air Ministry (in 1935, Büssing-NAG also built a 3-axle Ts with Flader body.

In 1941, the firm of A. Werner in Leverkusen built about 100 Ts 0.8 (TFL 8/41) on Opel 1.5-ton chassis, and in 1943 another hundred on 2-ton Peugeot chassis. The tank held 800 liters of a water-foam mix.

The Luftwaffe ordered numerous TFL 15 or TFL 15/43 (built 1942-43). The TFL 15 was mounted on the 3-ton uniform chassis made by Opel and the Mercedes L 3000 S as of the end of 1943, and sold for 17,000 RM. They proved themselves well in air raids: during the four air raids on Stuttgart on July 25-29, 1944, 31 TFL 15's were used successfully against 1984 major fires.

In 1943 the TFL 25 successor on Mercedes L 4500 A (all-wheel drive) was produced; 110 units were built. In all, some 750 TFL 25 trucks were built.

*Tanker-pumper of the Hannover Fire Police, photographed 3.24.1940.* Photo: Iveco-Magirus Archives.

*F.S. 145 tanker-pumper of the Remscheid Fire Police on Magirus L145 chassis, photographed 9/13/1939.* Photo: Iveco-Magirus Archives.

*Ts 2.5 a tanker-pumpers, ready for delivery to the air Ministry, photographed 2/25/1941.* Photo: Iveco-Magirus Archives.

*A Ts 2.5 (Henschel) fighting a fire. Date of photo unknown.* Photo: LFV Niedersachsen.

*Henschel tanker-pumper 2.5 a with TS trailer.* Photo: BF Archives, Göttingen.

*TS 25 tanker-pumper on DB chassis for the Luftwaffe.* Photo: Metz Archives.

# Hose Trucks and Tenders

*Hose tender with Metz body on MB L 3750 chassis, with enclosed hose reels, made for the Luftwaffe around 1937.* Photo: Metz Archives.

*Hose tender with pump trailer, built in 1939.* Photo: BF Archives, Göttingen.

*Heavy hose truck of the Frankfurt-am-Main Fire Police, on Mercedes-Benz LS 3000 chassis, 1938.*
Photo: MB Museum.

*Heavy hose truck on Mercedes-Benz L 3000 chassis, of the Berlin Fire Police, 1939.* Photo: MB Museum.

*SSK Heavy Hose Truck for the Aachen Fire Police.*

Photo: Iveco-Magirus Archives.

*Opel Blitz hose trucks of the Berlin Fire Department with pump trailers. The pictures were taken after 1945; note the Russian lettering on the trailer: Slusheniy furgon posharnaya sekcya (Service Truck, Fire Department).*

Photo: FMB.

*Hose truck on Magirus F.S.145 chassis, for the Air Ministry, photographed 10.20.1940.* *Photo: Iveco-Magirus Archives.*

*S 4.5 Hose Truck on Magirus FS 145, built in 1937. This truck was used by the Hannover BF until about 1961.* *Photo: BF Archives, Hannover.*

*Heavy hose truck on Klöckner-Deutz 3-ton chassis, for the Bremen Fire Police.* ***Photo: Iveco-Magirus Archives.***

*Hose truck (Metz body), able to carry some 2300 meters of hose.* ***Photo: Metz Archives.***

# Equipment Trucks

*Magirus F.S. 145 equipment truck for the Regensburg Fire Police.* Photo: Iveco-Magirus Archives.

*Hose truck on DB chassis, of the Berlin Fire Police, 1939.* Photo: FMB.

*Hose truck (Metz body) on MB LS 3750 chassis, of the Königsberg Fire Police.* Photo: Metz Archives.

*Magirus Crew and Equipment Truck, LGK-See, Kiel.* Photo: BF Archives, Göttingen.

*N 56 equipment truck of the Berlin BF after the war, with yellow front end and tarpaulin, in service until the mid-sixties.* Photo: FMB.

*Equipment truck on MB 8.5-ton chassis with 3-ton crane, in service in Berlin since 1928.* Photo: FMB.

*Equipment truck (3-ton crane) on Henschel chassis, of the Kassel Fire Police.* Photo: Metz Archives.

*An equipment truck of the Kassel Fire Police ready to tow a KS 25 of the SHD.* Photo: Metz Archives.

*Equipment truck on MB L 3750 chassis (Metz body with extended crane.* Photo: Metz Archives.

*Magirus M50 equipment truck of the Wuppertal Fire Police, with crane extended (in 1.5 minutes), photographed 7/21/1938.* Photo: Iveco-Magirus Archives.

*The first command vehicle of the Berlin BF in 1937, with extended antenna mast and loudspeaker. It was meant to be a central post for giving out information in cases of large-scale damage, and as a command post for firefighting.* Photo: FMB.

# Other Fire Vehicles

*"Radio cab with sending and receiving apparatus for wireless telegraphy and telephony" (original text in "Feuerlöschpolizei").* Photo: MB.

*A Magirus L30a crew truck for the Litzmannstadt (Lodz) Fire Department in the Warthe Region, photographed 1/4/1940.* Photo: Iveco-Magirus Archives.

*Unbelievable but true: an office truck of the Berlin Fire Police, on Opel P4 chassis.* Photo: FMB.

*An ambulance of the Greifswalde city government, on Hansa-Lloyd chassis, photographed 9/6/1937.* Photo: Iveco-Magirus Archives.

*Koebe tanker-pumper with 2-4000 liter tank capacity, on 3-ton chassis, with water cannon for foam operation. Photo: Koebe Firefighting Technology, Wenzel Publishing, Marburg.*

*A fire chief's car in "Beetle" form (Type 87). This car, not made in great numbers, was driven by, among others, higher fire police officials, such as Richard Bange of the Hannover FSchP. This 4-wheel-drive VW was built between 1941 and 1944.* *Photo: VW Museum Archives.*

*Tanker-pumper on Mercedes-Benz LGS 3000 chassis with Metz body for the "Reich Party Day Organization, Nürnberg." Emblem of the Police retouched after 1945.* Photo: Metz Archives

*A special fire truck (CO2 KS) on DB chassis, Tyle LG 3000 (1938-39) of the "Reich Party Day Organization, Nürnberg." The winch has a pulling power of 3500 kilograms. This "Märzfeld vehicle" was obtained in 1938.* Photo: DB Archives.

In 1939, there were 1.8 million FF members, organized in 50,000 units. In 1933, 500,000 more men were counted, but they were removed from the FF because of double duty (SA, SS, NSDAP, etc.). The military draft decreased the number of firemen further, but at the war's end there were still 1.1 million, including the HJ Fire Department members. In September 1933 the FF were reorganized into "Readiness Crews", each to have 70 men, divided into two platoons with three groups each. Every platoon had a heavy group (LF 15 and 4 SSK), the first groups had one LF 8 and four hose trucks. The FF constituted up to 40% of the entire firefighting strength in the cities.

*Two-piece firefighting unit, 1937-38: Metz aerial ladder on DB 3.5-ton chassis, Type LD 3000, and KS 25 on DB LS 3750 chassis, of the city of Giessen.* Photo: DB Archives.

*Three-piece firefighting unit (TFL, SDL and GLG) of the Wedding Fire Department (a district of Berlin).* Photo: FMB.

*Hermann Göring (extreme right, with white uniform) observes the firefighters at the Coulisse House of the State Theater, 7/30/1936.* Photo: FMB.

# Before the Storm

*Display of a Metz aerial ladder truck to Danish fire officials at the Berlin Fire Headquarters on Linden Street.* Photo: BAK 94/16/18a.

*KL 26 of the Dresden Fire Department: displayed on the Dresden Fairgrounds for the first "Red Rooster" exhibition in 1935. The mighty Almag-Hilpert front-end pumps are easy to see.* Photo: LFV Niedersachsen.

*Turned out for a parade are the eighteen fire trucks ordered from Metz in 1934.* Photo: LFV Niedersachsen.

*1939: The firefighters of an SHD group standing before a KzS 8 and reporting to the Troop Leader.* Photo: StA Mannheim.

*An unknown fire truck with Metz shutters and ladder! The IID license is for the Palatinate area.* Photo: Bartl/StA Speyer.

*November 10, 1939: extinguishing the fire at the synagogue in Brühl.* Photo: Neff/BPK.

*Two Magirus DL 26 ladder trucks with Fl 145 chassis, delivered to the Cologne Fire Police on March 2, 1939, in a public demonstration to appeal for winter welfare contributions.* Photo: BF Archives, Cologne.

*An LF 15 taking in water with suction hose.* Photo: Bartl/StA Speyer.

*Practice with the new Magirus ladder truck (Type K30) on the Burgstrasse grounds of the Fire Department in Frankfurt am Main, photographed 9/24/1936. The vehicle was delivered on 2/10/1935 and bore factory number 28833. Here the ladder is being raised.* Photo: LBSt. Hessen.

*At right, an old Magirus fire truck of about 1918, built in similar form until 1930, still in wartime service for the city of Mühlheim on the Ruhr. At left a Magirus ladder truck on Mercedes-Benz chassis of the Mühlheim BF.* Photo: StA Mühlheim/Ruhr.

## Chief Orpo, August 31, 1936:

"...shows that the Berlin Fire Department is the least well-equipped fire department of all professional firefighters in Germany. The age of the fire trucks can be described as legendary, and the speed of the trucks is fully insufficient for a large city."

The most recently purchased fire trucks were 21 years old and only reached 30-35 kph. These weaknesses became obvious when, on 7/30/1936, the coulisse house of the Berlin State Theater on Prince Friedrich Karl Street caught fire, and Senior Fire Director Wagner referred to the striking weaknesses in his report on 8/4/1936. The City Commissioner of Berlin, on 9/25/1936, ordered the immediate purchase of twelve fire engines, six mechanical ladders and one driving-school vehicle. The vehicles were put into service four to five months later.

In the first two war years, numerous old trucks that had been built during the Weimar Republic were still in use. If they had not yet been replaced by the new standard vehicle generation, they remained in service.

*An old Magirus fire engine likewise still performed its service for the Fire Police in Oldenburg.* Photo: LFV Niedersachsen.

***A fire truck of the 1st company of volunteer firemen in Speyer, photographed 6/26/1940, with open hose trailer, built in 1933.*** *Photo: Bertl/StA Speyer.*

*Winter 1939-40: A water main burst during construction of a bunker. A fire truck of the Saarbrücken Fire Police pumps the excavation out.* Photo: StA Saarbrücken.

*Winter 1939-40: An old fire truck has to be towed by horses in this cold winter!* Photo: StA Saarbrücken.

*Between 1933 and 1940, the Cologne Fire Police ordered five DL-26 trucks from Magirus, three of them built in 1939 on Magirus FL 145 chassis. Here one of these trucks is displayed on "Police Day" in Cologne.* Photo: BF Archives, Cologne.

*Fire officers with Dreyer oxygen masks climb the ladder.* Photo: AMF/ StA Bochum.

*July 1940: Equipping the SHD with vehicles. At right are LF 25 engines and Metz aerial ladder trucks, at left hose and equipment trucks, personnel trucks and chiefs' cars.* Photo: PKB.

*Long-nosed Magirus personnel trucks, Type 8 engines and aerial ladders of the Berlin Fire Department before the Olympic Stadium.* Photo: BAK 94/16/25A.

*Fire Police vehicles on display before the Olympic Stadium in Berlin. In the front row are chiefs' cars and courier cycles, behind them hose trucks, engines and ladder trucks.* Photo: BAK 94/16/23A

*The fire trucks march by the foreign guests before the Olympic Stadium in Berlin: LF 15, a DL 26 and another LF 15.* Photo: BAK 94/16/24A.

*Cross-country trip of a Cologne FE unit: first a KzS 8, then LF 25 (Mercedes L 3750 chassis).* Photo: BF Archives, Cologne.

*A Cologne company in their KzS 8 during an overland trip. Photo: BF Archives, Cologne.*

*A Cologne firefighting unit resting during an overland trip: nine KzS 8, one DL and six LF 25 (Mercedes L 3750 chassis) can be seen.*
*Photo: BF Archives, Cologne.*

*September 24, 1936: At the Burgstrasse Firehouse in Frankfurt am Main, the use of the front pump of a Magirus fire engine is demonstrated. Notice the sheet-metal panels on the left and right sides of the bumper.* Photo: LBSt. Hessen.

*A Magirus DL 30 + 2 on Magirus M 50 L chassis at the Burgstrasse Firehouse in Frankfurt am Main.* Photo: LBSt Hessen.

*Illustration: LFV Niedersachsen.*

# Fire Police, Forward!

During the Third Reich, the firefighters were organized along two lines: under the Reich Air Ministry and Reich Interior Ministry. On January 12, 1934 the concept of fire police was introduced, subordinate to the Interior Ministry. To be sure, Göring, as Commander of the Luftwaffe, could create his own firefighting units on the basis of the Air Raid Law of January 26, 1935. On November 28, 1938 the law pertaining to firefighting was enacted, the Minister of the Interior determined which communities should establish a fire police and which professional fire departments should be transferred to the fire police. Sixty-five fire police units were then established, and they recruited men from the professional firefighters. In Prussia, every city with more than 100,000 inhabitants had to have a professional fire department.

On May 4, 1937 the establishment of the SHD (Safety and Rescue Service) was ordered, to be set up locally; later it was formed into supraregional motorized units. The SHD was to see service in firefighting, clearing rubble, rescuing those trapped in rubble, repairing buildings, providing medical services, locating supplies and their contermination. At the end of 1939, the first SHD units (motorized) were set up, mainly with the KzS 8 on 1- or 1.5-ton Opel chassis. The equipping of the SHD in air-raid regions was directed by the Reich Minister of Air Travel: the Army Equipment Office delivered the vehicles to the communities, but they remained government property and bore WL license plates. Equipping with materials and vehicles was very extensive in comparison with the fire police. In May 1942 the SHD unit (mot.) became the LS unit (mot.) of the Luftwaffe; those units that were not taken over were designated LS Police. Several LS units (mot.) formed one LS regiment. The Air Ministry planned to set up eighty motorized LS units, but only twenty were approved. By the end of the war, 53 motorized Luftwaffe LS units existed, commanded by regimental staffs and stationed in the air districts.

SHD Unit 34 mot. was established in Nürnberg in 1940; it consisted of a staff, two firefighting and detoxifying teams, one repair team, one medical team with a patient transport echelon, one replacement echelon and a supply train. The unit had an ideal strength of 493 men (19 cars, 18 cycles, 75 trucks). On instructions from the RMdL, the SHD unit was assigned a third firefighting team without an E-platoon in September 1940 . . . all three teams received a light firefighting platoon as their third platoon.

An LS unit (mot.) was equipped with motor vehicles as follows:

12 KS 25, 6 KzS 6, 4 detoxifying vehicles, 2 detoxifying equipment vehicles, 4 KL 26, 4 SKW, 8 repair team vehicles, 8 repair equipment vehicles, 1 misfire vehicle, 1 air compressor, 1 RKW, 6 medical vehicles, 6 ambulances, 3 medical equipment vehicles, 1 telephone vehicle, 1 intelligence vehicle, 3 fuel vehicles, 5 field kitchen trucks, 6 distribution trucks, 1 repair shop truck, 1 workmen's truck, 3 reserve trucks, 20 motorcycles (8 with sidecars), 20 cars, total 127.

The SHD had 24 TSA and 20 KS at Nürnberg in March 1937. On April 1, 1941 the SHD had the following motor pool in Nürnberg:

79 cars, 37 motorcycles, 15 KS (ideally 36!), 18 KzS 8, 7 KL, 5 hose trucks, 1 intelligence vehicle, 214 light trucks, 6 sirens, 2 truck trailers (ideally 43!) and 49 motorized bicycles. Total: 333.

Ideal numbers of vehicles: 425 (Nürnberg), 416 at Mannheim (reached 70%), 473 at Frankfurt am Main (reached 78%).

# Locations of Fire Police (FSchP) Commands as of January 1941

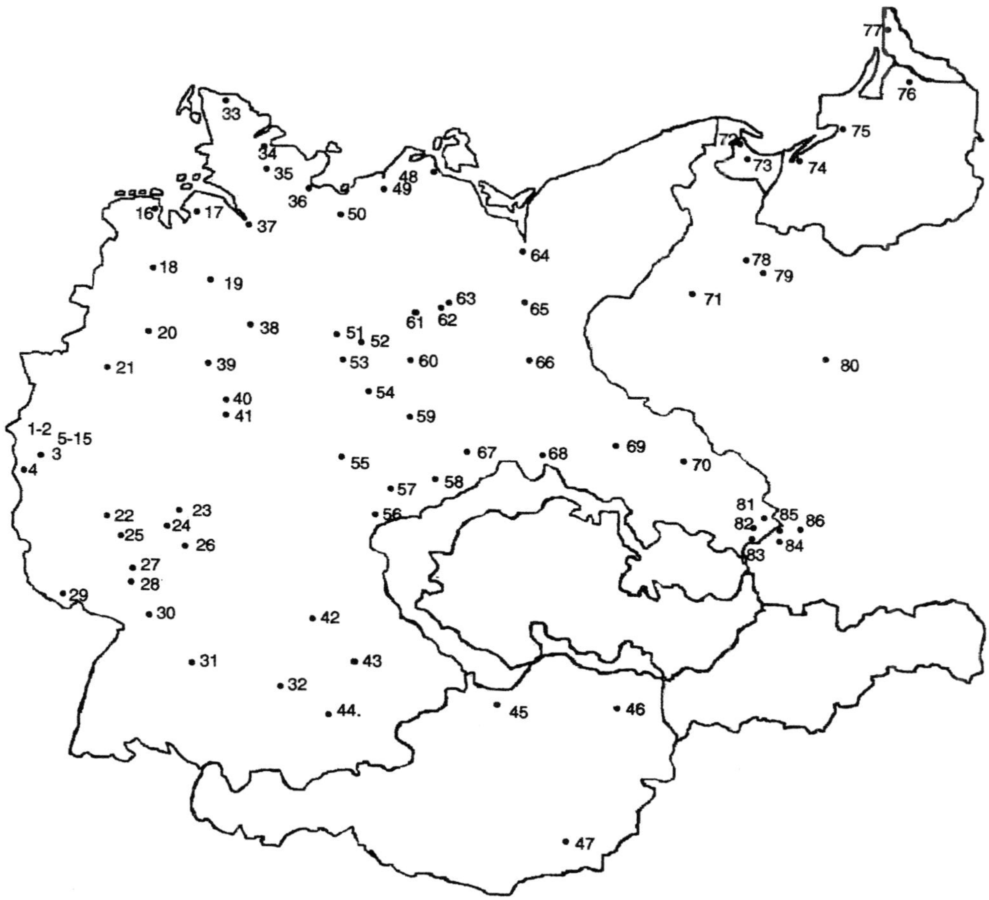

1. Mönchen-Gladbach
2. Düsseldorf
3. Köln
4. Aachen
5. Bochum
6. Dortmund
7. Duisburg
8. Krefeld-Uerdingen
9. Essen
10. Gelsenkirchen
11. Hagen
12. Mühlheim (Ruhr)
13. Oberhausen
14. Remscheid
15. Wuppertal
16. Wilhelmshaven
17. Bremerhaven
18. Oldenburg
19. Bremen
20. Osnabrück
21. Münster
22. Wiesbaden
23. Offenbach
24. Frankfurt (Main)
25. Mainz
26. Darmstadt
27. Ludwighafen (Rhein)
28. Mannheim
29. Saarbrücken
30. Karlsruhe
31. Stuttgart
32. Augsburg
33. Flensburg
34. Kiel
35. Neumünster
36. Lübeck
37. Hamburg
38. Hannover
39. Bielefeld
40. Wesermünde
41. Kassel
42. Nürnberg
43. Regensburg
44. München
45. Linz
46. Wien
47. Graz
48. Stralsund
49. Rostock
50. Schwerin
51. Braunschweig
52. Magdeburg
53. Halberstadt
54. Halle
55. Erfuhrt
56. Plauen
57. Zwickau
58. Chemnitz
59. Leipzig
60. Dessau
61. Brandenburg
62. Potsdam
63. Berlin
64. Stettin
65. Frankfurt (Oder)
66. Guben
67. Dresden
68. Görlitz
69. Liegnitz
70. Breslau
71. Posen
72. Gotenhafen
73. Danzig
74. Elbingen
75. Königsberg
76. Tilsit
77. Memel
78. Bromberg
79. Thorn
80. L tzmannstadt
81. Beuthen
82. Hindenburg
83. Gleiwitz
84. Kattowitz
85. Königshütte
86. Sosnowitz

# Polish Campaign and Occupation

On September 8, 1939, German Luftwaffe units appeared over Warsaw, which was surrounded by German troops. Ju-52 bombers, dive bombers and transports, fully loaded with incendiary bombs that were shoveled out the open side doors into the air over the city. 400 aircraft in all attacked the Polish capital that day in three or four flights. 560 tons of explosive bombs and 72 tons of incendiary bombs were dropped. All day long, 500,000 artillery shells of all calibers were fired into the city. Around noon, the column of smoke rose over Warsaw to a height of almost 4000 meters. There were fires all over the city, particularly in the central districts. For lack of water, almost no firefighting could be done. The fire trucks of the Polish professional firemen turned out: Fiat pumpers, Mercedes trucks and Magirus aerial ladders, built under license or bought in Germany before 1939, tried to extinguish the fires, like that in the flaming royal palace near the Old [p. 81] Market. There the firefighters had already been in action on September 18 to save the

*A 2.5-ton Fiat fire truck riddled by bullets.* Photo: LFV Niedersachsen.

*September 1939 in Warsaw: Pumpers and ladder trucks of the Polish firefighters in action on Marzalkov Street.* Photo: Batuk.

renowned ballroom. Ladder trucks were sent to the burning gas tanks in the suburb of Wola. Colonel Stanislav Gieysztor, Chief of the Warsaw Fire Department, and his more than 1100 firefighters fought in vain against the sea of flames. For the battle against the fire, 23 pumpers and ladder trucks of the Lodz Fire Department had arrived in Warsaw a short time before. Six vehicles from the Poznan Fire Department had also arrived.

After Warsaw had fallen and the Polish armed forces had capitulated, high-ranking officers of the Fire Police from the police headquarters inspected the destroyed Polish capital and questioned the Polish officers of the Warsaw Fire Department about their experiences in fighting the fires. Shortly before the war, Warsaw had had some 500 firefighters, including sixteen officers, but no volunteer firemen. Their equipment in terms of vehicles consisted in 1937 of 13 pumper trucks, four aerial ladder trucks, three tankers and five other pumpers. All of the aerial ladder trucks had been bought in Germany before the war, as had some of the other vehicles. In Poland itself, 2.5-ton Fiat trucks with front pumps and 200-liter water tanks had been built under license from Fiat. In September 1939 tha Warsaw Fire Department was enlarged to 1160 men. During the combat, more than 360 fires had been reported and fought on a single day. The losses of Polish firefighters are believed to have been 20 dead and 40 injured. A large number of their vehicles were destroyed or badly damaged.

*Members of the Fire Police of Section F at Orpo Headquarters examine the remaining fire trucks of the Polish firefighters in Warsaw and question the leading Polish fire officers about the fires caused by the air raids and ground attacks. Photo: LFV Niedersachsen.*

*Another Fiat fire truck (No. 311) that likewise shows signs of gunfire (bullet holes in the right door).* Photo: LFV Niedersachsen.

*A 2.5-ton Fiat fire truck with front pump, belonging to the Warsaw Fire Department in 1939, seen after the battle for the Polish capital ended. This truck had a 200-liter tank and was built under license in Poland. The vehicle (No. 411) shows signs of damage: bullet holes and the loss of the left rear dual wheel.* Photo: LFV Niedersachsen.

*Warsaw during the occupation. A Polish fire engine stands ready in the firehouse at 3 Chlodna Street.* Photo: BAK 135/818/4.

*Into action!* Photo: BAK 135/818/6.

*The Polish firefighters had the status of a "foreign auxiliary police" and were subordinated to the Orpo commander at Krakau. The vehicles, though, did not have Pol license plates, but OST ones.* Photo: BAK 135/818/7.

*A fire pumper (Mercedes-Benz chassis) on the way to a fire.* Photo: BAK 135/818/8.

*Fire Police trucks in Mönchen-Gladbach, 1940. Note the different Magirus emblems on the two trucks at left. From left to right: Magirus, a Magirus ladder truck (K 30L type with M40L chassis, serial number 905565, delivered 5/28/1938), a Magirus pumper, two KS 25 on Magirus chassis, LF 15, and chief's car.* Photo: StA Mönchen-Gladbach.

# Ready for Action in Germany

*Large-scale firefighting maneuvers in Mönchen-Gladbach in 1940. From left a KS 25, a KL 26 + 2 m, an ambulance and another KS 25.* Photo: StA Mönchen-Gladbach.

*An SHD unit with KzS 8 and TSA ready to march in Hannover. At left is Fire Chief Fritz Heimberg, the unit's chief instructor. The training platoon or "Musterlöschzug" had come into being.* Photo: LFV Niedersachsen.

*The firefighting platoon has arrived: one car, two pumpers, one ladder truck and one ambulance.* Photo: LFV Niedersachsen.

# Fire Police Regiments

*The Fire Police Regiment "Sachsen" assembled at Beeskow on April 12, 1940. Above (left to right, with 1944 ranks): Gen.Maj. of Police Walter Goldbach, Director, FSchP Office, Orpo HQ, Gen.Lt. Adolf von Bomhard (Chief, Command Office, Orpo HQ), Gen.Lt. of Police Johannes Meyer (Inspector General of Firefighting), Kurt Daluege (Chief, Orpo HQ) and Regiment Commander Gen.Maj. of*

After the experiences of the Polish campaign in September 1939, Fire Police regiments after the French model were organized beginning in November 1939. They were to follow the combat troops and fight fires that resulted from combat action or protect important places (refineries, tank farms, etc.). The first to be formed was FSchP Regiment 1 "Sachsen" (Commander Hans Rumpf), which saw service in the West and then in Russia (Group East) in 1941, with the task of protecting the petroleum fields of Baku. Parts of the regiment were then transferred to Romania to protect the oil fields and refineries in Ploesti. In January 1941, the FSchP Regiment 2 "Hannover" was founded, and in the summer and November of 1941, under the leadership of the East Prussian "Firemen's General" Alfred Fiedler, FSchP Regiments 3 (Ostpreussen) and 4 (Ukraine). In 1942 came FSchP Regiments 5 (Bohemia and Moravia) and 6 (Netherlands). The motor pool of Regiment 5 included, in all, 2500 motor vehicles, many of them from foreign lands. The regiments were disbanded in the first half of 1943, and the units were utilized as independent mobile intervention reserves, principally in parts of Germany endangered by air raids.

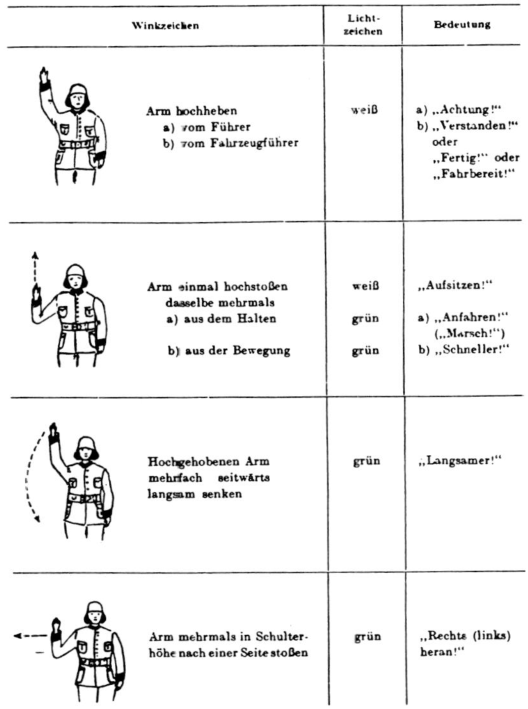

| Winkzeichen | Licht-zeichen | Bedeutung |
|---|---|---|
| Arm hochheben<br>a) vom Führer<br>b) vom Fahrzeugführer | weiß | a) „Achtung!“<br>b) „Verstanden!“ oder „Fertig!“ oder „Fahrbereit!“ |
| Arm einmal hochstoßen | weiß | „Aufsitzen!“ |
| dasselbe mehrmals<br>a) aus dem Halten | grün | a) „Anfahren!“ („Marsch!“) |
| b) aus der Bewegung | grün | b) „Schneller!“ |
| Hochgehobenen Arm mehrfach seitwärts langsam senken | grün | „Langsamer!“ |
| Arm mehrmals in Schulterhöhe nach einer Seite stoßen | grün | „Rechts (links) heran!“ |

*Hand signals and instructions for fire truck drivers.* *Illustration: Heimberger/Fuchs.*

*A bus of FSchP Regiment Sachsen.* *Photo: Joop.*

*Lined up behind their LF 8 with TSA.* *Photo: Joop.*

*On the march with FSchP Regiment Sachsen.* *Photo: Joop.*

*LF 15 of FSchP Regiment Sachsen.* *Photo: Joop.*

*A firefighting group lines up behind their LF 15. Note the shutters on the rear of the truck.* Photo: StA Mühlheim/ Ruhr.

# In Germany Until 1941

*A KS 175 of SHD Platoon 119 of the 54th Unit in Berlin.* Photo: FMB.

*Fire Police line up in front of the firehouse: at left an LF 15 with front pump on Krupp chassis, then a KzS 8 with TSA, also on Krupp chassis, at right a Luftwaffe LF 25 and another KzS 8.* Photo: StA Mühlheim/ Ruhr.

*Berlin 1941: An SHD platoon lines up with Opel Blits trucks of the detoxifying service.* Photo: BAK 94/16/ 33A.

*The F. & B. Troop 186 of the Weissensee Guard in 1939, with an open fire truck powered by wood gas.* Photo: FMB.

*An overloaded KzS 8 with trailer in Berlin at the beginning of the war.* Photo: Dietz Collection.

*KzS 8 trucks with trailers in Berlin at the beginning of the war.* Photo: Dietz Collection.

*A firefighting group of the SHD with a KzS 8 in Berlin.* Photo: Dietz Collection.

*A KzS 8 is inspected. The Mercedes chassis were delivered to city fire departments, the Opel chassis to the Reich's own units.* Photo: Paulitz/BF Leipzig.

*A firefighting group lines up behind their KzS 8 and TSA.* Photo: Dietz Collection.

*Mönchen-Gladbach, February 1942: the Magirus ladder truck (grille covered with blankets) removes icicles from roofs.* *Photo: StA Mönchen-Gladbach.*

*There were "peacetime fires" in wartime too: Fire in the freight section of the Anhalt Station in Berlin, which spread to several offices and the roof (12/30/1940). Several ladder trucks are in use.* *Photo: LBSt. Berlin.*

*The FSchP of Kassel celebrating its fiftieth anniversary in 1941.* Photo: Carl Eberth/BF Kassel.

*A DL 26 of the Hagen Fire Police.* Photo: Metz Archives.

*In this fire at Bonn City Hall on February 10, 1942, three people lost their lives: a locksmith and two SHD members. The picture shows a Luftwaffe LDL in action.* Photo: StA Bonn.

*A KzS 8 of the Stuttgart Fire Police on Mercedes-Benz L 1500 chassis.* Photo: Ebel.

*An LF 8 of the Fire Police with a TSA trailer.* Photo: Johanssen.

*An LF 8 on DB L 1500 chassis.* Photo: Johanssen.

*The same vehicle in Hamburg in the "Gomorrha" Exhibition in 1993.* Photo: Gerlach.

*LF 15 of 1938, with 3-ton Opel Blitz chassis and Koebe body, of the Nordenham Fire Police, photographed in February 1994.* Photo: AMF.

KLÖCKNER
MAGIRUS
Das neue
schwere LÖSCHGRUPPENFAHRZEUG
MAGIRUS
SLG
ut nach den Vorschriften des Reichsführers ϟϟ und Chefs der deutschen Polizei, gemäß
sterial-Erl. v. 16.2.1940, genehmigt v. Generalbevollmächtigten für das Kraftfahrwesen.
LÖCKNER-HUMBOLDT-DEUTZ AG
WERK ULM / MAGIRUS-FEUERWEHRGERÄTEBAU
ULM/DONAU
ZIALFABRIK FÜR FEUERLÖSCHFAHRZEUGE UND -GERÄTE ALLER ART

DIESEL
MERCEDES-BENZ-METZ
Kraftfahr-Spritzen

MAGIRUS
JERWEHR-
OSS-
HRZEUGE

MAGIRUS
TRANSPORTWAGEN
FÜR KLEINMOTORSPRITZEN
Goliath
MAGIRUS

*An LF 25 of 1938 with Koebe body, now owned by the Göttingen Fire Department. Left: the cab. Below: the body maker's name stamped into the bottom.* Photos: AMF.

In 1838 the Koebe firm began to produce the body for an LF 25 for the RLM. Nothing is known of its use in 1939-45. On June 22, 1942 the truck was registered as Pol 76820 in Göttingen. It was never painted green and kept its original red color after 1945.

After the war, the LF 25 was rebuilt into an equipment truck and one window was covered. Later the LF 25 was used as a driving-school vehicle. In 1968 the LF 25 was taken out of service.

Technical Data:

| | |
|---|---|
| Vehicle type | GLG/LF 25 |
| License numbers | HIL 14-528 |
| | HA 11-6032 |
| | BN 62-5036 |
| | GÖ 2057 |
| Chassis type | DB L 3750 |
| Chassis number | 306 426438 |
| Engine number | 67 503 29 |
| Engine power | 100 HP |
| Weight empty | 6050 kg |
| Allowable weight | 9000 kg |
| Front axle load | 3000 kg |
| Rear axle load | 6000 kg |
| Length overall | 7900 mm |
| Width overall | 2100 mm |
| Height overall | 2600 mm |
| Tires | 8, 25-20 |
| Front tire pressure | 4.25 atm. |
| Rear tire pressure | 4.75 atm. |
| Fuel capacity | 120 liters |
| Pump: | |
| Manufacturer | Amag Hilpert, Nürnberg |
| Type | JF IV/2500/2-stage |
| Number | 770128/37 |
| Year built | 1937 |
| Power | 2500 l/min. |
| Body: | |
| Manufacturer | Koebe, Lückenwalde |
| Number | 9902 |
| Tank capacity | 400 liters |

*KS 25 on Magirus FS 145 "long-nose" chassis.* Photo: Johanssen.

*A KS 15 with Büssing chassis (Burglöwe 350 type) and Flader body, circa 1935.* Photo: Johanssen.

*A KS 25 on MAN chassis, Type D 1, with Fischer body (1936).* Photo: Johanssen.

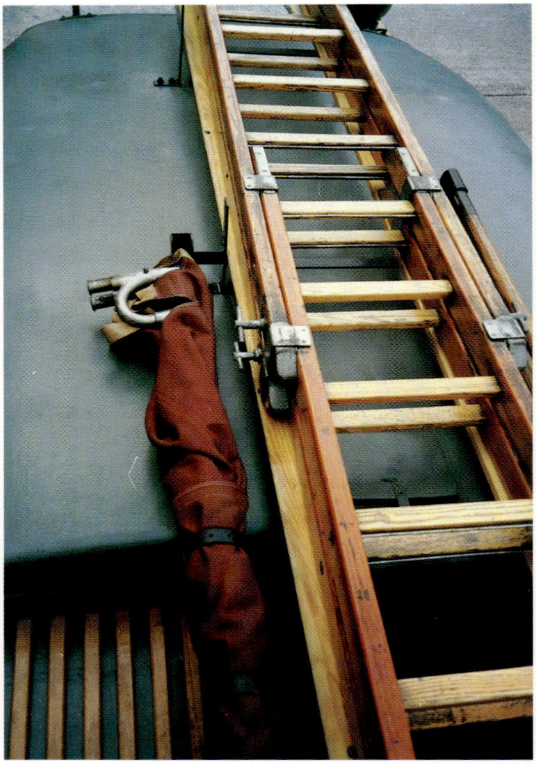

*The cab of an LF 8, finished in wood.* Photo: Dietz Collection.

*The roof of an LF 8: at left the folding stretcher, then the extension ladder and the pike.* Photo: Dietz collection.

*Wooden seats for the crew of an LF 8.* Photo: AMF.

*KS 15 with Magirus M 27 chassis, built in 1938, used by the Oberndorf volunteers, out of service in 1978.* Photo: Johanssen.

*An LF 15 on Mercedes-Benz L 3000 F chassis, built in 1942, today at the Mercedes-Benz Museum.* Photo: Johanssen.

*An LF 15 of 1940 by Magirus (FS 145 chassis), built for the Munich Fire Police.* Photo: Johanssen.

*A DL 24+2 of the Celle volunteers. Ordered 4/12/1938, it was delivered 8/30/38 (serial number 905580) with M30L chassis. The steel ladder had to be turned by hand, and the price was 26,553 RM. The vehicle is now at the Danneberg/Tramm Fire Muesum.* Photo: AMF.

*This DL 46 of the Berlin Fire Department remained in service until the mid-sixties.* Photo: FMB.

*Magirus DL 26 of the Cologne Fire Police in action.* Photo: BF Archives, Cologne.

*A KS 25 with front pump, used by the Luftwaffe in Crete and taken over by the Greek firefighters after 1945. Taken out of service around 1980.* Photo: Ebel.

*One of the rare LF 8's in original paint. The sand-colored version of the LLG with a sand filter for the engine was made for Rommel's Afrika-Korps and is now kept at Autohaus Hövels in the Sauerland.* Photo: Gras Collection.

*Front view of the LF 8 with body by Flader, of the Ahlem/Hannover volunteers.* Photo: AMF.

*DL 26 on MB chassis, with Diesel engine, ordered by the RLM, today owned by a collector in Oldenburg.* Photo: AMF.

*A DL 27 (Metz ladder on Büssing NAG Type 500 chassis), built in 1938 for the Braunschweig Fire Department.* Photo: Johanssen.

*Luftwaffe fire trucks on the march in winter; in front a heavy hose truck, followed by two Metz ladders on Mercedes-Benz chassis and one LF 25.* Photo: BF Archives, Cologne.

*After an RAF air raid on Cologne, the firefighters attack the burning houses.* Photo: BF Archives, Cologne.

*Firefighting in Kassel on October 23, 1943, after a British air raid destroyed much of the city. Trucks of the local Fire Police and the Luftwaffe on Frederick Square, looking toward Upper King Street.* Photo: Unken/StA Kassel.

*Taking on water at the Karlsaue in Kassel, on the shore of the Little Fulda, on 10/23/1943. The burned-out Orangerie is shown, with the Marble Bath beside it. At left is an LF 25 of the Luftwaffe, in the background an LF 15 of the fire police, at right another of the same type, as well as Luftwaffe buses to transport the men.* Photo: Unken.StA Kassel.

# Not A Chance . . .

*A messenger lost in the fire, Hamburg, July 28, 1943.* *Photo: Brunswig/PKB.*

On September 11, 1944 Darmstadt was destroyed. 12,000 people lost their lives, 70,000 their homes. The 150 firefighters of the city stood helpless before the sea of flames. Klaus Schmidt writes in "Die Brandnacht":

"On the Autobahn, firefighters and rescue troops from the whole area between Mannheim, Frankfurt am Main and Mainz gathered between two and three o'clock. Up to 3000 men with 220 fire pumpers were there by 6:00. Waiting for orders, they had to watch as the city went up in flames. Pushing forward into the center of that storm of flames was impossible. So now they do what is possible on the edge of the fiery circle."

*Ruined houses in Hamburg.* *Photo: LBSt. Hamburg.*

*Extensive areas burn near Hamburg Harbor.* *Photo: LBSt. Hamburg.*

*Not a chance: one of the major oil fires of World War II in Hamburg.* Photo: LBSt. Hamburg.

*No help is possible: a girl's body in Hamburg, July 28, 1943.* Photo: Brunswig/PKB.

# IN THE WEST

Units of FSchP Regiment 1 reached the West in 1940, going to Holland and France. The French firefighting regiment in Paris was subordinated to Commander Rumpf. The regiment's main areas of action were the French Atlantic harbors and their fuel tanks, which were now being used by the German Navy. British commandos and the RAF were able to set the tanks afire again and again: in Brest, Le Havre, St. Nazaire, St. Lorient.

The Luftwaffe also sent its firefighters to France: LS units (such as No. 34) and numerous vehicles and crews came to protect the important military airfields during the Battle of Britain or to protect the harbors in which the transport ships stood ready for "Operation Sealion."

*July 6, 1940: In front of Fire Headquarters in Paris, an LF 8, an LF 15 (Magirus) and another LF 15 (Mercedes-Benz) await orders to march.* Photo: Helmerding.

*A unit of FSchP Regiment Sachsen in action in Paris, 1940.* Photo: Helmerding.

*Vehicles of Unit 1, FSchP Regiment 1, get ready to leave Paris for their third action in Le Havre on July 16, 1940.* *Photo: BAK 85/44/23 (above), BAK 85/44/22 (below).*

*Luftwaffe fire trucks in action after an air raid on Cologne.* Photo: Paulnitz/BF Cologne.

*An LF 25 taking on water.*
Photo: Paulnitz/BF Cologne.

*Vehicles of an SHD unit (mot.) of the Luftwaffe in the area around Le Havre, June 1940. In the center are two DL-26 on Mercedes-Benz LS 3750 chassis.* Photo: BAK LOC 1689.

*Three-ton Opel Blitz box trucks of an SHD unit in the Le Havre area, June 1940.* Photo: BAK LOC 1688.

*A pump at work in Le Havre.* Photo: BAK 94/16/6A.

*A KS 25 in action, Le Havre, 1940.* Photo: BAK 94/16/8A.

*A KS 25 in action at a tank fire, Le Havre, 1940.* Photo: BAK 94/16/7A.

*An LF 15 of FSchP Regiment 1 in action.* Photo: Helmerding.

*A ladder truck in front of FSchP Regiment 1 headquarters in Paris.* Photo: Helmerding.

*The "goulash cannon" was also part of the motor pool. This one bears the emblem of the Police, belonged to the 5th Company, FSchP Regiment 1 "Sachsen", and was photographed in France in 1941.* Photo: Joop.

*A cobbled-up kitchen of FSchP Regiment 1 "Sachsen" in France, 1941. At first, equipped with a smokestack, it annoyed Commander Rumpf, who could not stand such "gypsy wagons." So the firemen created an apparatus allowing them to retract the smokestack.* Photo: Joop.

*France, 1941: A captured Renault of the 5th Company, FSchP Regiment 1.*

Photo: Joop.

*An FlTs 2.5 of an LS unit (mot.) at Cambrais, France, in 1941.* Photo: Böning.

*The same LS unit after applying foam from their FlTs 2.5. In the center is a soldier in an asbestos suit.* Photo: Böning.

*Loading a truck in Brest.* Photo: Helmerding.

*An LF 8 and a three-axle truck loaded on flatcars.* Photo:Joop.

*The bus driver is directed onto the flatcar. Photo: Helmerding. Loading a bus.* Photo: Helmerding.

*Car, trailer and LF 8 are already aboard in Brest. Photo: Helmerding.*

*Trucks and tankers of the 5th Company, FSchP Regiment 1 are transported from Brest to Krakau by train.* Photo: Joop.

# IN THE EAST
## Russia

The Russian action of the "Group East" (two companies) had the long-range goal of protecting the oil fields in the Caucasus. The unit, though, was only able to make its way as far as Kharkov. The climatic conditions and road conditions set clear limits for the fire trucks. Almost all the captured cars that had been lined up in the motor parks at Dunkirk and Brest broke down. The fire trucks' inability to drive cross-country led to further losses. Many vehicles had only 12 cm ground clearance; they had to be dragged through the mud by towing tractors or teams of men (sometimes covering six kilometers a day!), and in winter they came to grief on the frozen, rutted airstrips. The towbars of the TSA trailers broke, and temperatures of 30 degrees below zero made firefighting, when it was possible at all, an ordeal. In November 1941 the "Group East" was withdrawn from the Eastern Front.

*Medium uniform personnel cars (command cars) in Galicia in July 1944. High Berlin police and RFSS officials inspected the hard-hit units in the East. Note the SS emblem on the right fender of the car in front.* Photo: Joop.

# The Burning of Kiev, 1940

As the Red Army withdrew, Kiev was mined at Stalin's orders, the mines exploded by radio and the city set afire. The Group East of FSchP Regiment "Sachsen" went into action against the burning city on September 15, 1941, under Fire Police Major Haase. The firefighting began on September 20. The time bombs left by the Soviets exploded at intervals, strong winds and ammunition of all kinds stored all over the city fanned the spreading flames. Since the canal was destroyed, water had to be brought from the Dniepr, 1.5 kilometers away, by hose lines. Two engineer battalions were assigned to firefighting; they blew holes in house fronts to stop the fire. Only after four days could the great fire be brought under control.

At the booty collection point, a fire broke out, growing to a major blaze that spread to the captured ammunition dump. At the same time, fires broke out in a building housing a store and apartments. The fires joined together in a short time and spread quickly. The fire trucks in a circle of some 30 km from Kiev had been abandoned by the Soviets. The water works had been blown up, so that the hydrants gave no water. Hose was requested by air from Krakau, but it never arrived. The use of Kiev as winter quarters for the troops was questionable. In the night of September 27-28, 1941 the fires drew back considerably and finally died out altogether.

*Mud in the East, autumn 1941. Teams of men have to tow the vehicles.* Photo: Joop.

*The retreat westward from Kharkov, November 1941. In front is a heavy hose truck, behind it KS 25 and LF 15.* Photo: Joop.

*In Dniepropetrovsk at 40 below zero: After being transported by rail in 1942-43, these FlTs 2.5's are winterized and moved to the largest supply airport of Stalingrad: "Dniepropetrovsk South."* Photo: Böning.

*Hose tender of an LS unit (mot.) on Mercedes-Benz chassis in Russia.* Photo: Böning.

*A hose tender with its crew in retreat at the Black Sea in 1944 (note the "baggage" carried on the truck).* Photo: Böning.

*TS 2.5 on three-axle Henschel chassis at a military airfield in Russia.* Photo: BAK 389/1073/28.

*A Henschel fire truck (Water Truck Kfz. 94) of the army on the Eastern Front.* Photo: BAK 137/1040/18a.

# Service in Germany

*Night action of the Fire Police in Berlin.* Photo: Ullstein Photo Service.

*Bomb damage in Berlin, 1940. On the street are several Fire Police vehicles, including one LF 15.* *Photo: Ullstein Photo Service.*

*An LF 25 (Mercedes-Benz L 3750 chassis) being ferried across the Rhine. Note the number III-25 between the two right-side doors.* *Photo: BF Archives, Cologne.*

*Action in Bremerhaven in March 16, 1941. A ship's boy took revenge by starting a fire aboard the luxury liner Bremen, which was lost along with a camouflaged ship. A Magirus aerial ladder truck erected its ladder to the funnel.* Photo: Paulitz/BF Hamburg

*A KS 25 of the SHD before the burned-out ocean giant.* Photo: Paulitz/BF Hamburg.

*KS 25 on Magirus chassis before the smoking ruins of the Bremen.* Photo: Paulitz/BF Hamburg.

*On April 10, 1941 the Berlin Fire Police battled a large fire in the Berlin State Opera after an RAF air raid the previous night (attack from 0:27 to 3:38 o'clock). The State Opera was struck by several incendiary bombs, but much of it could be saved. Five ladders are erected against the building. At the left is a command vehicle, then two LF 25 and at right a KzS 8.* Photo: Ullstein Photo Service.

*Unter den Linden: an LF 25, an LF 15 and two KzS 8 stand before the Berlin State Opera. Ladders are up on this side of the building too, carrying hose to the firefighters on the roof.* Photo: Ullstein Photo Service.

## DNB Report of April 10, 1941

In their attack on the capital of the Reich in the night of April 9-10, the British were helped by the extraordinarily clear view in the bright moonlight, which let them recognize all their targets even from great heights. They dropped explosive and, particularly, incendiary bombs, mainly on such buildings as had a reputation in the world's culture. Among others, the State Opera and the State Library were badly damaged.

*In front of the State Opera is an LF 25 with its hose connected to a hydrant.* Photo: FMB.

## Report of the Berlin Police Chief, April 10, 1941

In this air raid, some fifty explosive and 4000 incendiary bombs were dropped. The attack left eleven dead and 74 injured. The Berlin Police Chief wrote on April 10 about the bombing of the State Opera at Unter den Linden 7: "Hit by a number of incendiary bombs. Theater building burned out. Stage area also damaged. Severe property damage resulted. Fire extinguished by fire police. On the square west of the Opera, 30 incendiary bombs extinguished. One dead, a theater guard killed by smoke. One injured (also by smoke) and taken to hospital."

*Rescued musical instruments from the Berlin State Opera on the street, 4/10/1941. In the background are fire trucks, including a ladder truck, after the fire was extinguished.* Photo: LBSt. Berlin No. 275294.

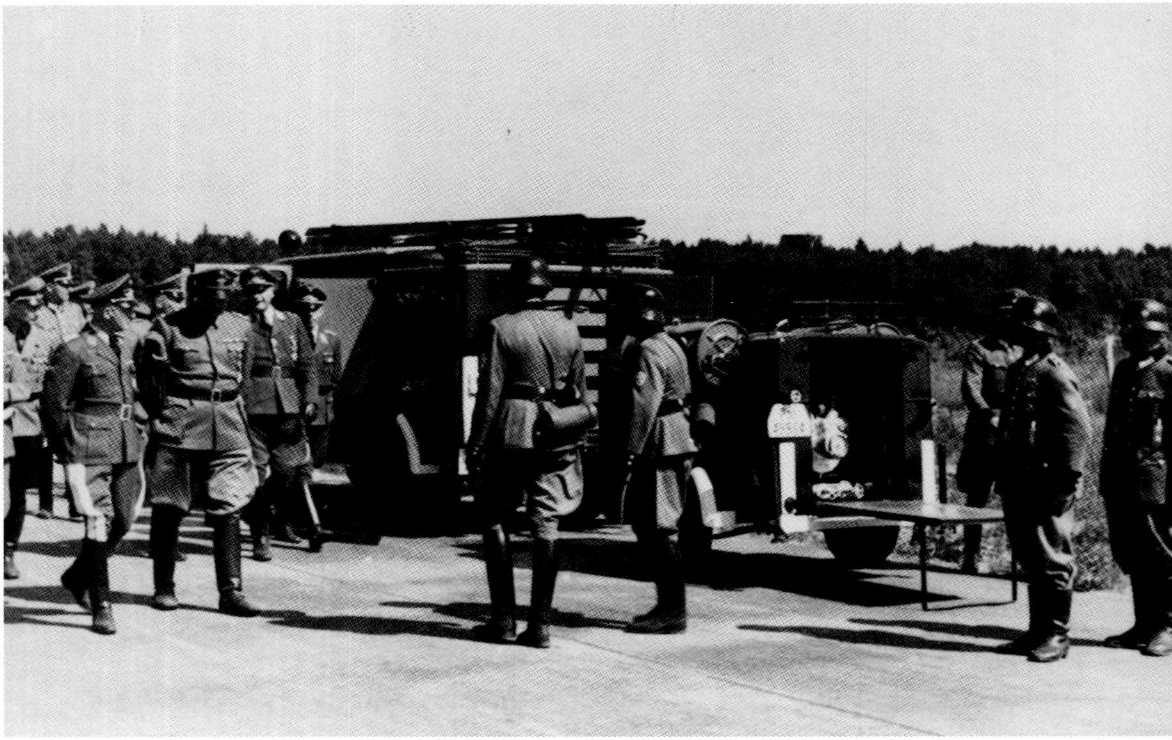

*Daluege is briefed by officers of the Luftwaffe and Fire Police. At the curb is an LF 8 with TSA.* Photo: BAK 94/16/10A.

*Ten LF 8 with Mercedes chassis and Metz bodies, II. Ast.* Photo: BAK 85/44/11

*Hoses are laid toward the fire.* Photo: BAK 94/16/11A.

*Practice: Hoses are unrolled . . . II. Ast.* Photo: BAK 85/44/12.

*. . . and hooked up to the trailers.* Photo: BAK 85/44/13.

*After the practice, the hoses are rolled up again. In the background are two LF 8's with trailers. The pumps have been lifted out of the trailers.* Photo: BAK 94/16/15A.

*At the end of the practice by FSchP Regiment 2, Daluege stands in his car to review the passing fire trucks.* Photo: BAK 94/16/14A.

*Unit 1, Fschp Regiment 2 laying foam.An LF 25 takes on water.* Photo: BAK 85/44/18.

*Practice at Neumünster, June 1941: LF 8 with TSA is prepared for service.* Photo: BAK 85/44/16.

*Pumpers and ladder trucks in practice on the Berlin-Stettin Autobahn.* Photo: FMB.

*The Celle Volunteer Firemen picked up an LF 15 in Ulm in 1942, one of three. A short rest beside the Autobahn.* Photo: Hische Collection.

**From the Report of the Commander of the Fschp Hamburg, Oberst of the Fire Police Dr. Zaps, on the air raids in the summer of 1943, September 1943**

"Maintenance of vehicles."

To make sure vehicles are kept in service, after the experiences of the catastrophe, the following requirements are to be established:

a) New fire vehicles with built-in pumps must be broken in properly by being driven at least 1000 km. The saving of fuel is only an apparent one, for the breakdown of a motor vehicle often takes place in the critical time while fighting a fire and cannot be excused by any economical measure. The unavoidable severe. engine damage today requires weeks and months of being out of service because of the difficulties of obtaining any spare parts. The recommendation cited in directions for cutting costs, of breaking in vehicles by occasional service trips, is rather pointless for fire vehicles, since utilization for transport purposes is impossible because of the body types. For fire vehicles, in the interests of readiness for service, an exception must be allowed. In the same sense, this also applies to portable pumps.

b) The supplying of spare parts is of no less significance. It must be striven for that large-scale repair shops of the firms or contract repair shops must maintain ample supplies of spare parts, also for firefighting apparatus—at least for standard vehicles, more or less corresponding to the divisions of the defensive zone . . . In the present rush for as high fire vehicle production as possible, one must not forget the maintenance of those already built!"

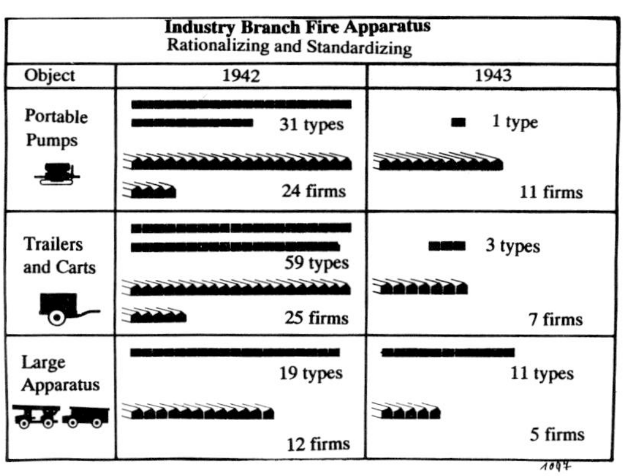


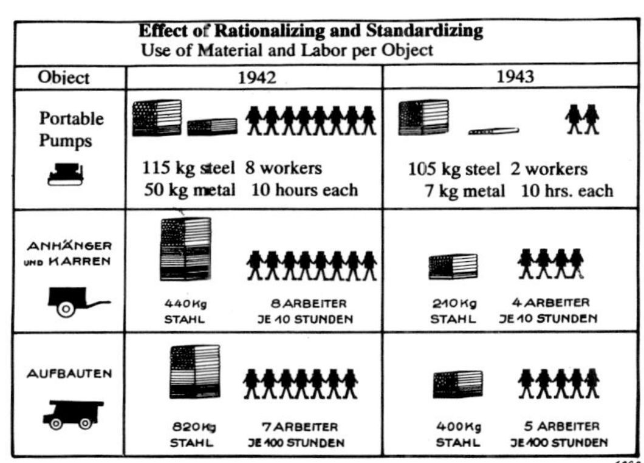


*Illustrations: Iveco-Magirus Archives.*

## H. Brunswig on Service Experience of Firefighters in World War II

According to Brunswig, the weak point of all fire vehicles was the clutch. Inexperienced drivers, driving over rubble, night drives in columns with frequent delays often caused the vehicles to break down on their first trip. If they were towed away by other fire vehicles, the latter themselves sometimes broke down from overstraining the clutch in starting up. This was all the more serious when spare parts could never be obtained and damage to the clutch meant two days at a repair shop. Just in June 1944, according to Brunswig, 22 defective clutches of Hamburg fire vehicles were removed!

*Two Luftwaffe LF 25 trucks stand ready, Berlin 1943.* Photo: FMB.

*Two pumpers of the Reichenberg Fire Department, Berlin.* Photo: FMB.

# Readiness!

*A firefighting platoon with two KS 25 and one DL.* Photo: Dietz Collection.

*An FE firefighting unit from Nauen waits for a guide's directions in Berlin in 1943.* Photo: FMB.

*Three KS 25 trucks in front of a firehouse.* Photo: Dietz Collection.

*A Luftwaffe ladder truck (Metz) on Martin Strèet in Cologne on May 31, 1942, just before the facade of a burned-out house collapsed. In front is a DL 26 ladder on MB Lo 3750 chassis.* Photo: IBF Archives, Cologne.

*Photo: Paulitz/BF Cologne.*

*An LF 25 of the 3rd F. & E. Team after taking a bomb hit in the inner courtyard of the fairgrounds on the Auenweg, Cologne, 1942: eleven dead. The truck had come to the fairgrounds on its own because a fire had been spotted there.* Photo: Paulitz/BF Cologne.

*A KS 25 on Magirus M45 chassis of the Fschp, 1943-44.* Photo: Paulitz/BF Cologne.

*KS 25's and ladder trucks. The GLG are built on various types of Magirus and Mercedes-Benz chassis.*
Photo: Paulitz/BF Cologne.

*KzS 8 on Opel chassis taking on water from the Rhine.* Photo: BF Archives, Cologne.

*Heavy 4.5 hose truck. with an LF 25 in the background.* Photo: Paulitz/BF Hamburg.

*S 4.5 heavy hose truck of the Hamburg Fire Police. Note the water-carrier symbol ("Hummel") on the rear of the truck. In the background is an LF 8 with trailer.* Photo: Paulitz/BF Hamburg.

# Drought and Flood

*Fire police vehicles in action during the flood catastrophe in the Ruhr area on May 18, 1943.* Photo: StA Bochum.

*LF 15 and LF 25 in action.*
*Photo: StA Bochum.*

*An LF 25 (Magirus), an old pumper, an LF 25 (Mercedes) and an LF 15.*
Photo: StA Bochum.

*December 1943: A woman driver for the Reichenbach Fire Department.* Photo: Schwahn/BAK 94/16/27A.

**From RDSS Directive No. 1/39 of June 28, 1939**

1. The Youth Leader of the German Reich makes available to the Reichsführer SS and Chief of the German Police in all air-raid shelters of the first and second class, according to extent of local need, assistance from the units of the Hitler Youth. The age of the boys to be commandeered for firefighting is set at fifteen years, whereby an even distribution among the three oldest years of the Hitler Youth members is to be observed. Their height should, if possible, not be less than 165 cm.

5. The firefighting division of the units is in groups and platoons (2 groups = 1 platoon). The largest unit (Gefolgschaft) that can be set up within the parameters of this organization shall, in a first- or second-class air-raid shelter, for the time being not exceed 80 boys, or in exceptional cases 100 boys.

8. The leadership of firefighting training is in the hands of the commander of the local fire police or the local fire chief, who is to work out a plan in collaboration with the local leader or local deputy of the Hitler Youth . . .

In an RFSS directive of April 21, 1943, the communities were required to solicit volunteers from women and girls between 17 and 40 years of age to fill the staffs of fire departments of communities of under 1000 inhabitants. Women and girls should, for example, be recruited as drivers of fire vehicles. Since large-scale volunteering did not occur, recruiting took place. In Constance in the autumn of 1944, 40 women were recruited by the fire department, and 27 joined the service. In all, 275,000 women reportedly joined the ranks of fire departments in the Reich.

*Two groups of women firefighters being trained. In the background is an LF 25, with an LF 8 and trailer in front of it.* Photo: Schwahn/BAK 94/16/32A.

*A water line to draw water from a brook is coupled.* Photo: BAK 78/81/1.

*It's still fun: Hitler Youths, under the direction of a fire chief, practice in Stahnsdorf near Berlin, September 1939.* Photo: Ullstein Photo Service.

*Fully exhausted Hitler Youths take a break after an air raid in Berlin late in 1943.* Photo: Ullstein Photo Service.

*Hitler Youth firefighters practice in the yard of the Düsseldorf Fire Department on Hütten Street in 1943. A KS 25 is being prepared for service.* Photo: StA Düsseldorf.

*A Hitler Youth fast command in service after an air raid in Düsseldorf on August 25, 1943.* Photo: Genzler/BAK 94/16/22A.

The "Völkische Beobachter" reported on May 1, 1936 that, in Bad Reichenhall in November 1935, "Germany's first Hitler Youth fire department" had been established. In December 1939 it was determined that the personnel for training in firefighting had been made available from the units of the Scouting Service. In 1940 a Hitler Youth fire brigade, with an average of 45 boys, was formed from the Hitler Youth Scouting Service. The RFSS ordered on March 31, 1941 that the firefighting reserves from the Hitler Youth should serve to strengthen the Fire Police and the fire departments.

The training of the Hitler Youth firefighters was regulated by a directive from the RFSS on August 15, 1939, in collaboration with the Reich Youth Leader. In the entire Reich area, 300 youth fire brigades were to be organized. In all, 300,000 members of the Hitler Youth are believed to have been called into firefighting service during the war.

*Members of the Fire Police and the Hitler Youth Fire Brigade on Ludwig Street in Munich after a heavy air raid in the summer of 1944. The Ludwig Church is burning in the background.* Photo: Photo Service, South German Publishers.

During the course of the war, as personnel losses increased, more and more Ukrainians were recruited for the Fire Police as of 1943. The Fire Police units gained a fourth company, which consisted completely of Ukrainians. Numerous Fire Police commands in the cities also accepted Poles, Ukrainians and ethnic Germans: in Hamburg, Braunschweig, Stuttgart, Munich, Dresden, Kassel, Nürnberg and Berlin. Foreigners were also accepted in factory fire departments. In a directive from Orpo HQ on March [illegible], 1944, it was stated that the percentage of foreign members of a factory fire department could not be more than half of the entire membership. "As long as Poles and East Europeans are recruited for factory fire departments, the duty, now as before, of wearing the prescribed "P" or "Ost" symbol remains. This symbol is therefore to be attached to their uniform (clothing)."

*Establishment of the "Ostpreussen" Firefighting Regiment, in the summer of 1941, by the "Fireman General" Alfred Fiedler. Recruiting regulations were circumvented. Photo taken June*

*Training of new firefighting units. Four LF 25's are in the background.* Photo: Behrens/BF Kassel

*In the background is a ladder truck of the Oberhausen Fire Police, at right an old ladder in Mühlheim Street after an air raid on June 15, 1943.* Photo: StA Oberhausen.

*A KzS 8 of a motorized LS unit after an air raid on Oberhausen, seen on Market Street on April 27, 1943.* Photo: StA Oberhausen.

*KS 25 taking on water.* Photo: Dietz Collection.

*Two KS 25 taking on water.* Photo: Dietz Collection.

# Bombing War 1943-1945

*Two KS 25 in a firefighting demonstration in Berlin. Two other trucks and a detoxification van have also arrived.* Photo: Dietz Collection.

*A group from Essen at the City Hall fire on Adolf Hitler Street in Bochum, 1943. Note the tactical symbol on the back of the truck.* Photo: StA Bochum.

*After the air raid on Wuppertal, an LS unit (mot.) cleans up on Völkling Street.* Photo: StA Bochum.

# Alarm in the Ruhr Region

*A KzS 8 during the first major attack on Dortmund on May 5, 1943, before the burning Wandelhalle and Probstei Church.* Photo: StA Bochum.

*Firemen rest after a major air raid on Duisburg on May 13, 1943.* Photo: StA Bochum.

The series of major air raids on Hamburg by the RAF began on Sunday, July 25, 1943: 741 bombers dropped 2300 tons of bombs on the city. The Commander of the Hamburg Fire Police, Oberst of the Fschp Dr. Zaps, directed approximately 9300 firefighters in 1943, divided into 30 divisions with three firefighting platoons each. The firefighting organization had 305 fire trucks and 935 trailers available: the FE Service had 180 fire trucks, the detoxifying units had 43 vehicles, the strengthened FD had 65, the factory air-raid teams 65 and the administrations had two. By July 26, 1943 the FE Service had lost 8 dead, 11 missing, 21 seriously injured and 61 slightly injured men. Eight fire trucks were fully destroyed, 27 others were damaged, and three ladder trucks, four hose trucks, one car and 20 motorcycles were also destroyed or damaged. On July 26, 1943, 122 Flying Fortresses of the 8th USAAF made a daylight attack on Hamburg. On the 27th, the following losses were counted: five dead, 17 missing, 17 badly and 32 slightly injured. Thirteen fire trucks were destroyed and twelve damaged. One ladder truck, two hose trucks, seven cars and 19 motorcycles were also lost, either totally destroyed or requiring long-term repairs. Of the thirty FE divisions in the LS area of Hamburg, only 22 were capable of service on June 27, 1943. On the night of July 27-28, 1943, 739 RAF bombers attacked Hamburg again and caused one of the most terrible firestorms of the whole war. A short time later, they were replaced by 54 four-engine bombers of the USAAF, which appeared over Hamburg and continued the destruction of the city. The firefighters' losses as of July 30, 1943 numbered 15 dead, six missing, six badly and two slightly injured, four fire trucks destroyed and five damaged. Two ladder trucks, two hose trucks, five cars and seven motorcycles were destroyed. After these two attacks, only 21 FE divisions were capable of service. The last attack was made on August 3, 1943, with 425 RAF bombers carrying 939 tons of bombs. The losses of the Fire Police this time were comparatively slight: one dead, two badly and six slightly injured; one fire truck was damaged.

Fahrzeugverteilung des FE-Dienstes

| Schlauchkraftw. | Kraftfahrleiter | Löschzug: Löschgruppe b | Löschzug: Löschgruppe a | Löschzug: Löschgruppe b | Löschzug: Löschgruppe a | Löschzug: Löschgruppe b | Löschzug: Löschgruppe a | Bereitschaften | FE-Abteilungen | LS-Gruppen |
|---|---|---|---|---|---|---|---|---|---|---|
| Skw.-099 | KL-4.251 | Kz-094 | KS-3.403 | KS-118 | KS-117 | Kz-295 | 5.251 Fr. +8.251 Fr. (Lokstedt) | 1/I | I | West |
| Skw.-257 | KL-301 | KS-3.254 | Kz-126 | KS-3.252 | KS-921 | 5.406 Fr. +8.552 Fr. | 5.252 Fr. +8.391 Fr. (Niendorf) | 2/I | | |
| Skw.-258 | KL-4.364 | KS-714 | KS-3.384 | KS-713 | KS-463 | Kz-092 | Kz-091 | 1/II | II | |
| Skw.-100 | KL-4.271 | KS-665 | KS-3.301 | 5.283 +8.301 | Kz-093 | Kz-573 | 5.407 Fr. (Rissen) +8.404 Fr. (Lurup) | 2/II | | |
| Skw.-155 | KL-4.361 | KS-517 | KS-3.281 | KS-515 | KS-679 | KS-926 | KS-925 | 1/III | III | |
| Skw.-259 | KL 46-033 | KS-119 | KS-465 | Kz-098 | Kz-096 | Kz-102 | Kz-101 | 2/III | | |
| Skw.-7.251 | KL-4.371 | KS-109 | KS-3.381 | KS-3.291 | KS-3.292 | Kz-104 | Kz-106 | 1/VII | VII | |
| Skw.-260 | KL 26-255 | KS-113 | KS-111 | KS-680 | KS-112 | KS-927 | Kz-097 | 2/VII | | |
| Skw.-158 | KL-4.252 | KS-3.341 | KS-3.282 Fr. (Blankenese) | KS-3.131 | KS-3.231 Fr. (Eidelstedt) | KS-914 | Kz-110 | 1/IX | IX | |
| Skw.-261 | KL-4.281 Fr. (Eidelstedt) | KS-512 | KS-511 | KS-916 | KS-915 | 5.415 Fr. (Osdorf) +8.393 Fr. (Groß Flottbek) | 5.271 Fr. +8.264 Fr. (Blankenese) | 2/IX | | |
| Skw.-101 | KL-4.362 | KS-3.361 | KS-3.385 | KS-107 | KS-642 | 5.408 Fr. +8.263 Fr. (Billwärder) | Kz-297 | 1/IV | IV | Ost |
| Skw.-262 | KL-4.382 | KS-3.395 | KS-3.392 | KS-712 | KS-634 | KS-928 | KS-924 | 2/IV | | |
| Skw.-157 | KL-4.351 | KS-3.461 | KS-3.391 | KS-666 | KS-519 | KS-918 | KS-917 | 1/V | V | |
| Skw.-167 | KL-305 | KS-3.391 Fr. (Bergedorf); 5.418 Fr. (Moorfleth) +8.253 Fr. | KS-3.261 Fr. (Bergedorf); 5.413 Fr. +8.255 Fr. (Spadenland) | KS-3.281 Fr. (Neuengamme); 5.282 Fr. +8.269 Fr. (Allermöhe) | KS-3.392 Fr. (Nettelnburg); 5.402 Fr. +8.267 Fr. (Reitbrook) | 5.321 Fr. +8.231 Fr. (Curslak); Kz-134 +8.273 Fr. | Kz-128 +8.2611 Fr.; 5.301 Fr. +8.252 Fr. (Kirchwerder Nord) | 2/V, 3/V | | |
| Skw.-7.252 | KL-4.363 | KS-3.412 | KS-3.413 | KS-640 | KS-464 | Kz-567 | Kz-572 | 1/VI | VI | |
| Skw.-264 | KL 26-284 | KS-462 | KS-461 | KS-518 | KS-923 | 5.401 Fr. +8.381 Fr. (Hummelsbüttel) | KS-3.351 | 2/VI | | |
| Skw.-156 | KL-4.282 Fr. | KS-633 | KS-3.401 Fr. (Wandsbek) | KS-3.393 Fr. (Wellingsbüttel) | KS-678 | Kz-109 | Kz-108 | 1/X | X | |
| Skw.-168 | | KS-3.251 Fr. (Wandsbek) | KS-922 | 5.381 Fr. +8.261 Fr. (Bramfeld) | 5.254 Fr. +8.403 Fr. (Rahlstedt) | 5.341 Fr. +8.402 Fr. (Berne) | 5.382 Fr. +8.262 Fr. (Farmsen) | 2/X | | |
| Skw.-265 | | 5.414 Fr. (Oldenfelde) +8.254 Fr. | KS-3.252 Fr. (Billstedt) | 5.392 Fr. +8.382 Fr. | 5.372 Fr. +8.268 Fr. (Volksdorf) | 5.417 Fr. +8.282 Fr. (Poppenbüttel) | KS-930 | 3/X | | |
| Skw.-266 | KL-4.341 | KS-3.411 | KS-3.404 | KS-667 | KS-516 | Kz-571 | Kz-289 | 1/A | A | Hafen |
| Skw.-134 | KL-4.253 | KS-514 | KS-513 | KS-669 | KS-668 | 5.412 Fr. +8.397 Fr. | KS-108 | 2/A | | |
| Skw.-267 | KL-303 | KS-636 | KS-630 | KS-671 | KS-670 | KS-3.253 | Kz-561 +8.254 | 1/B | B | |
| Skw.-153 | KL-4.261 | KS-3.382 | KS-114 | KS-673 | KS-672 | Kz-127 | Kz-291 +8.394 Fr. (Wilhelmsburg) | 2/B | | |
| Skw.-135 | KL-4.221 | KS-631 | KS-3.402 | KS-710 | KS-632 | Kz-559 +8.265 Fr. (Moorwerder) | Kz-568 | 3/B | | |
| Skw.-104 | KL-4.342 | KS-3.383 | KS-3.393 | KS-675 | KS-674 | Kz-124 | Kz-105 | 1/C | C | |
| Skw.-136 | | KS-106 | KS-105 | KS-677 | KS-676 | 5.391 Fr. +8.256 Fr. (Finkenwerder) | Kz-103 | 2/C | | |
| Skw.-137 | KL-4.361 | KS-635 | KS-3.394 | KS-644 | KS-643 | Kz-125 +8.311 Fr. | Kz-300 | 1/VIII | VIII | Ab. VIII |
| Skw.-268 | | KS-711 | KS-3.272 | KS-931 | KS-919 | Kz-294 | Kz-290 | 2/VIII | | |
| Skw.-154 | K[illegible]04 | KS-639 | KS-641 | 5.283 Fr. +8.281 Fr. | 5.404 Fr. +8.291 Fr. | 5.416 Fr. +8.395 Fr. | 5.411 Fr. +8.396 Fr. | 3/VIII | | |

*Fire vehicles of the FE Service in Hamburg.* Illustration: StA Hamburg.

*A Magirus KL 26 ladder truck of the SHD in action after an air raid in Hamburg.* Photo: LBSt. Hamburg.

*LF 25 fighting a fire in Hamburg.* Photo. Gerlach Collection.

*A ladder truck in action after an air raid in Hamburg.* Photo: LBSt. Hamburg.

Slight damage to vehicles, particularly tire damage, was taken care of on the spot. As of August 5, 1943, over 400 tires were patched at the main firehouse alone.

After the heavy air raids in the summer of 1943, the vehicle losses as of August 3 were as follows:

| | **Total Losses** | | | | | | **Damaged, long-term repairs** | | | | | |
|---|---|---|---|---|---|---|---|---|---|---|---|---|
| | LF 8 | LF 15 | LF 25 | DL | S | TS | LF 8 | LF 15 | LF 25 | DL | S | TS |
| WL | 0 | | 8 | 2 | 5 | 4 | 3 | - | 8 | 0 | | 10 |
| Gem. | 5 | 5 | 7 | 1 | 0 | | 4 | 22 | 4 | 7 | | 8 |
| Total | 5 | 5 | 15 | 3 | 5 | 4 | 7 | 22 | 12 | 7 | | 18 |

According to Brunswig, the FE Service had lost a total of 67 dead and 147 injured during the summer's air raids, and 160 vehicles had been lost. The losses included: 75 fire pumpers, 21 detoxifying trucks, 6 ladder trucks, 8 hose trucks, 15 cars and 48 motorcycles. According to the official report of the Commander of the Hamburg Fire Police, Dr. Zaps, 27,033 people had been killed by September 14. This was one of the greatest catastrophes of the war.

*A Magirus ladder truck fighting a fire in Hamburg.* Photo: Paulitz/BF Hamburg.

*A Luftwaffe LF 25 in action after an air raid.* Photo: Paulitz/BF Hamburg.

*LF 8 in action in Hamburg. Note the lettering on the pump trailer. Aerial ladders are raised in the background.* Photo: Paulitz/BF Hamburg.

**Firefighting Forces in the Summer Air Raids on Hamburg, 1943**

| | FF-Ber | FF-Gr. | Fschp Units Comp. | LS-Pol. Grps. Units | LS-Abt. mot. Comp. | Units VFB | Other Units | WLS ES NSDAP |
|---|---|---|---|---|---|---|---|---|
| 7/25 | 16 | 89 | 6 | 18 | 10 Co. | 30 Ber.<br>55 Gr.VFD | 5.5 Ber.<br>10 VFD | 640<br>200 |
| 7/26 | 2 | 19 | 4 | 8 | 12 Co. | 30 Ber. | 1 Ber. | - |
| 7/27 | 2 | 20 | 4 | 3 | 12 Co. | 30 Ber. | 2 Ber. | - |
| 7/28 | 11 | 30 | 6 | 27 | 16 Co.<br>5 Ber. | 30 Ber.<br>55 Gr.VFD | 8 Ber.<br>4 Ptn.<br>10 Gr.VFD | - |
| 7/29 | 2 | 3 | 2 | 10 | 6 Co.<br>5 Ber. | 30 Ber. | 4 Ber. | - |
| 7/30 | 23 | 45 | 1 | 14 | 20 Co.<br>5 Ber. | 30 Ber.<br>65 VFD | 2 Ber. | - |
| 7/31 | 2 | 13 | 1 | 1 | 14 Co.<br>5 Ber. | 30 Ber. | 2 Ber. | - |
| 8/1 | - | - | 1 | - | 7 Co. | 30 Ber. | 1 Ber. | - |
| 8/2 | - | - | 1 | - | 8 Co. | 30 Ber. | 1 Ber. | - |
| 8/3 | - | - | 1 | 4 | 4 Co. | 30 Ber. | 7 Ber. | - |

**Structure of an FE Unit (ideal) as ordered January 25, 1940**

| Unit Staff | 1st FE-Bereit. | 2nd FE-Bereit. | 3rd FE-Bereit. |
|---|---|---|---|
| | (all like 1st FE-Bereit.) | | |

FE-Bereit.
Light Platoon 1st Fire Group 1/8 1 KzS 8; 2nd Fire Group same
Light Platoon same
Heavy Platoon s. Fire Group 1/8 1 KS 15/25, 2nd Heavy Group same
Detoxifying Group 1/8+1 Entg. Truck, 2nd E Group same

*Aerial ladders in action in Hamburg. The fire truck in the background has cutters on its roof. Note the varying hose connections and sizes.* Photo: Paulitz/BF Hamburg.

*Firefighting after an air raid on Hamburg. At left is an LF 25.* Photo: Paulitz/BF Hamburg.

*An LF 25 of the Fire Police on Lützower Street, Bremen, December 1943.* Photo: StA Bremen.

*An LF 25 of the Fire Police fighting a fire on North Street in Bremen, December 1943, after the 122nd air raid.* Photo: StA Bremen.

*Burned-out fire trucks of FE Bereitschaft 2/X on the Steindamm in Hamburg after the air raid of July 28, 1943.* Photo: Paulitz/BF Hamburg.

*A destroyed fire truck at Firehouse 9 in Elmsbüttel after the air raid of July 25, 1943. Picture taken July 27, 1943.* Photo: PKP/Brunswig.

*A Luftwaffe ladder truck after an air raid on Cologne.* Photo: BF Archives, Cologne.

*An LF 25 of the Cologne Fire Police.* *Photo: BF Archives, Cologne.*

*A Luftwaffe Metz DL 26 (Mercedes L 3750 chassis) in action near the main railway station in Cologne.* *Photo: BF Archives, Cologne.*

*A DL 26 and an LF 25 fighting a fire in Hamburg.* Photo: Paulitz/BF Hamburg.

*Two Magirus DL 26 ladder trucks on M45 chassis in action in Hamburg.* Photo: Paulitz/BF Hamburg.

*An equipment truck winches a dead horse out of a canal on February 18, 1941.* Photo: Paulitz/BF Hamburg.

*An equipment truck towing a heavy hose truck of Fschp Regiment 2.* Photo: Paulitz/BF Hamburg.

*September 20, 1943: Rescuing safes of jewels from the city lending office from the ruins of the building in the Trankgasse.* Photo:Paulitz/ BF Hamburg.

*A Klöckner-Deutz LF 15 loaded on a flatcar.* Photo: Dietz Collection.

*For lack of fuel, fire trucks could not always be driven by road to burning cities. Here LF 15 and LF 25 trucks are being transported by rail.* Photo: Dietz Collection.

*Wiesbaden Fire Police vehicles in action after an air raid.* Photo: Paulitz/BF Wiesbaden.

*An LF 25 in action after an air raid on Munich in 1944.* Photo: South German Publishing Photo Office.

*An open fire truck in action.* *Photo: Paulitz/BF Düsseldorf.*

*Falling pieces of stone have damaged a Luftwaffe Metz ladder truck in Cologne.* *Photo: BF Archives, Cologne.*

*Firefighting at Hamburg Harbor near the Fish Market. A KzS 8 with trailer in the foreground.* **Photo: Paulitz/BF Hamburg.**

*A Luftwaffe KS 25 with Magirus M 45 chassis near the Elbe, Hamburg Harbor.* **Photo: Paulitz/BF Hamburg.**

*A damaged ladder truck in Lübeck, 1942.* **Photo: Paulitz/BF Hamburg.**

*The front end of a KL 46.* **Photo: Paulitz/BF Hamburg.**

*A KzS 8 of the Luftwaffe SHD and an LF 15 of the Hamburg Fire Police on Klöckner-Deutz chassis.* **Photo: Paulitz/BF Hamburg.**

*A pump trailer in firefighting action on Schelling Street, Munich after an air raid in March 1943.* Photo: StA Munich.

*An LF 15 fighting a fire in Munich after 217 bombers had dropped 56 tons of bombs on March 10, 1943.* Photo: StA Munich.

*An aerial ladder in action in Munich after an air raid.* Photo: StA Munich.

*After an air raid on March 10, 1943, firemen take a break and are given food and drink. In the background is an old fire truck of the "Deutsche Sprengchemie Werke Geretsried."* Photo: StA Munich.

*September 14, 1942: On Ostertor Street in Bremen, an aerial ladder is in action at police headquarters.* Photo: W. Cüppers/StA Bremen.

*After the 103rd air raid on Bremen on September 14, 1942, SHD vehicles are seen on General Ludendorff Street. Note the hose lines on both sides of the street.* Photo: W. Cüppers/StA Bremen.

*After the 92nd air raid on Bremen on June 26, 1942: the scene at the Sielwall intersection the next morning.* Photo: W. Cüppers/StA Bremen.

*An LF 25 fighting a fire on Hospital Street on October 20, 1943.* Photo: City Historical Museum, Leipzig.

*An LF 25 fighting a fire in Leipzig on October 20, 1943.* Photo: City Historical Museum Leipzig.

*An LF 25 taking on water from the Elster, April 6, 1945.*

***Photo: City Historical Museum, Leipzig.***

*Liebfrauenberg in Frankfurt am Main, 1943: LF 25 of the Fire Police and Luftwaffe (at right) taking on water from a man-made fire pond to which water is brought through hose lines.* ***Photo: Essinger/IfStaG Frankfurt am Main.***

*October 1943, Hanauer Landstrasse, Frankfurt am Main: Heavy hose trucks of a motorized LS unit in action after an air raid.* Photo: Göllner/IfStaG Frankfurt am Main.

*Ostend Street, Frankfurt am Main, October 1943: An Opel Blitz 1.5-ton, an LF 15, and a ladder truck in action. At right front is a hose cart with messenger's bicycle and spray nozzle.* Photo: Göllner/IfStaG Frankfurt am Main.

*An aerial ladder in action on Kaiser Street, Frankfurt am Main, October 1943.* Photo:Schmidter/IfStaG Frankfurt am Main.

*The Voigt & Höffner Factory on Hanauer Landstrasse, Frankfurt am Main, date unknown: an LF 25 fighting a fire after an air raid.* Photo: IfStaG Frankfurt am Main.

*Vehicles of the Fire Police in Mühlheim/Ruhr before a high bunker. The front KzS 8 bears no police emblem, although its license plate (Pol) shows it to be a police vehicle.* Photo: StA Mühlheim/Ruhr.

*A completely destroyed KzS 8 on Opel Blitz chassis after an air raid in Mühlheim/Ruhr, probably after the RAF air raid on June 23, 1943.* Photo: StA Mühlehim/Ruhr.

*After an American daylight air raid on Braunschweig on August 5, 1944: a man collapses of smoke inhalation on Julius Street.* Photo: StA Braunschweig.

*Cyriak-Ring, Braunschweig, August 5, 1944: At right is an LF 8 of the Fire Police, at left an LF 25 from Wolfsburg (note wolf emblem behind door).* Photo: StA Braunschweig.

*Numerous fire trucks gather at the Neustadt Depot in Bremen, October 1944. Among others, we see two 4.5 heavy hose trucks, four LF 8 with trailers, one LF 15, various cars and a light ladder truck.*

***Photo: StA Bremen.***

*A firefighting group of the LS Police goes into action with gas masks on Papen Street, Bremen, October 1944. Behind the firemen is a KzS 8.* Photo: StA Bremen.

*An LF 25 of the Bremen Fire Police taking on water, September 1944.* Photo: StA Bremen.

Because of large-scale fires in the hearts of the old cities, the "Wassergasse" (water lane) tactic was developed. In 1943 and 1944, especially in Bremen and Braunschweig, it helped to rescue tens of thousands of people, including 6000 in Braunschweig who had tried to reach the air-raid shelter and had been surrounded by fire in the Old City. The firemen fought their way along one street and extinguished or cooled the burning house fronts on both sides of the street, to a distance of about 100 meters. These "water lanes" were extended to the air-raid shelter and the people were taken to safety through them.

*Several LF 25 trucks in Bremen after the 138th air raid, October 1944. Photo: StA Bremen.*

*A firefighting group drives through the ruins of Bremen (North Street, 1945). Photo: StA Bremen.*

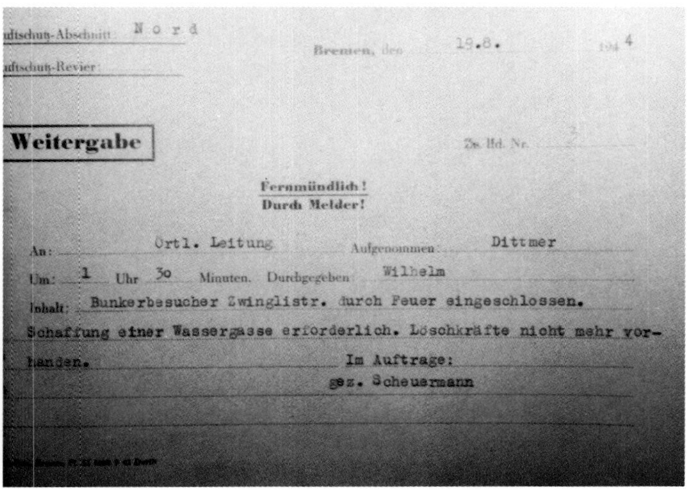

...uftschutz-Abschnitt: Nord

...uftschutz-Revier:

Bremen, den 19.8. 1944

**Weitergabe**

Zw. lfd. Nr.

Fernmündlich!
Durch Melder!

An: Örtl. Leitung Aufgenommen: Dittmer

Um: 1 Uhr 30 Minuten. Durchgegeben: Wilhelm

Inhalt: Bunkerbesucher Zwinglistr. durch Feuer eingeschlossen. Schaffung einer Wassergasse erforderlich. Löschkräfte nicht mehr vorhanden.

Im Auftrage:
gez. Scheuermann

*A report on the creation of a water lane. Photo: StA Bremen.*

In 1944 and 1945, Nürnberg, the "City of Reich Party Days", was the target of many major RAF air raids. On September 10, 1944 MAN was hit hard and its truck production was ended. The Faun works lost 100% of their production capacity. The losses of fire vehicles increased ominously in 1944: in the attack on April 6, 1944, seven LF 25, two LF 8, three S 3 hose trucks and one ladder truck were lost. On January 2, 1945, 1000 four-engine bombers attacked Nürnberg, and the LS Police alone lost 29 dead and 88 wounded. Firefighting teams came from Würzburg, Dresden, Leipzig, Bamberg, Bayreuth, Coburg and Regensburg. In the double attack on February 20 and 21, 1945, ten LSP and FF members died and 26 were wounded. On duty were ten FE teams from the LS area of Nürnberg-Fürth, 36 FF, 10 reserve teams of the 15-kilometer fire assistance zone, nine FE teams from the Ansbach and Regensburg commands, three companies of Fire Police Unit 8 (Air-Raid Regiment 3), and the motorized Air-Raid Regiments 22, 26, 28, 37, 38 and 44, 26 army firefighting groups, ten from the railroad system and three from military airports, plus 700 men of the TN. In this attack, 1356 people were killed and almost 70,000 became homeless. The fire vehicle losses of the firefighters were high, since the men were surprised by a third wave of bombers while they fought fires.

The 2nd Company of FSchP Unit (mot.) 8, which was underway from Ebermannstadt to Nürnberg on February 20, 1945, suffered high vehicle losses to low-flying plane attacks on Route 4; six vehicles were so badly damaged that they were unusable. One tanker and one engine were burned out completely.

At the end of March, Nürnberg fire vehicles were attacked by low-flying planes while fighting fires. Bullets tore through their bodies and engine blocks.

*Symbol of LS Unit (mot.) 34.* Illustration: AMF.

*Hose lines bring water to Tafelfeld Street to fight fires on October 19, 1943.* Photo: StA Nürnberg.

A DL 32 of an LS unit (mot.) with protective covering on the ladder, in front of St. Sebald's Church in Nürnberg, October 3, 1944. Photo: StA Nürnberg.

A Luftwaffe aerial ladder on Karoline Street in Nürnberg, August 28, 1943. Photo: StA NÜrnberg.

*An LF 15 of the Nürnberg Fire Police in front of the building at Kohlengasse 2, March 9, 1943.* Photo: StA Nürnberg.

*An LF 15 of the Nürnberg Fire Police taking on water at the Schuckert Square pond, October 19, 1944.*
Photo: StA Nürnberg.

*After an air raid on the Kreuzberg section of Berlin, street-clearing operations were carried out. On the left side of Zimmer Street are two tower trucks with side ladders; at the right an aerial ladder truck.* *Photo: LBSt Berlin.*

## New Tactics and Training

*A ladder truck in action on Kirschner Street in Halle on February 27, 1945, after an air raid.* *Photo: StA Halle.*

*Practice had to continue during the war, here with the nozzle of a Magirus ladder truck of the Berlin Fire Police.* Photo: AMF/FBM.

*SHD hoselaying practice in Cologne. In the background is an LF 25 (Mercedes-Benz L 3750 chassis).* Photo: BF Archives, Cologne.

### *Seating Order in the LF 15 or 25 Fire Truck*

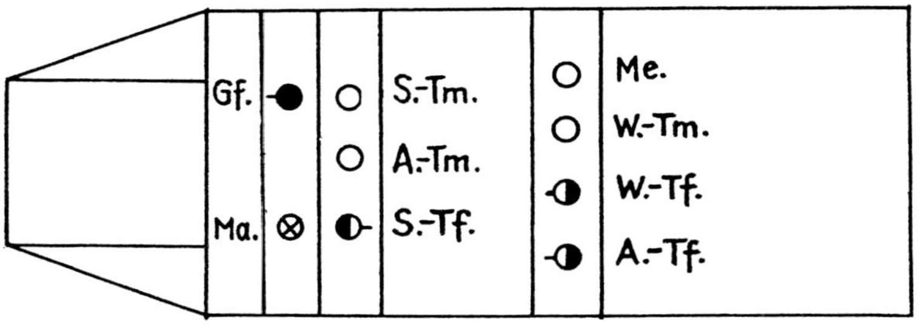


*Picture 15 / Seating Order*)[credit]* *Illustration: Heimberg-Fuchs*

**Command: "March!"**

The fire truck drives away.

**Command: "Halt!"**

The command "H-a-l-t!" should be drawn out.

"Halt!" can also be given by signal.

**Command: "Off!"**

The crew gets off and stands behind the vehicle as in pictures 11 and 12.

In alarm service the giving of commands "To the truck!", "On!" and "Off!" can be omitted.

*A Luftwaffe firefighting crew lined up behind their LF 25, Bochum 1943.* *Photo: AMF/StA Bochum.*

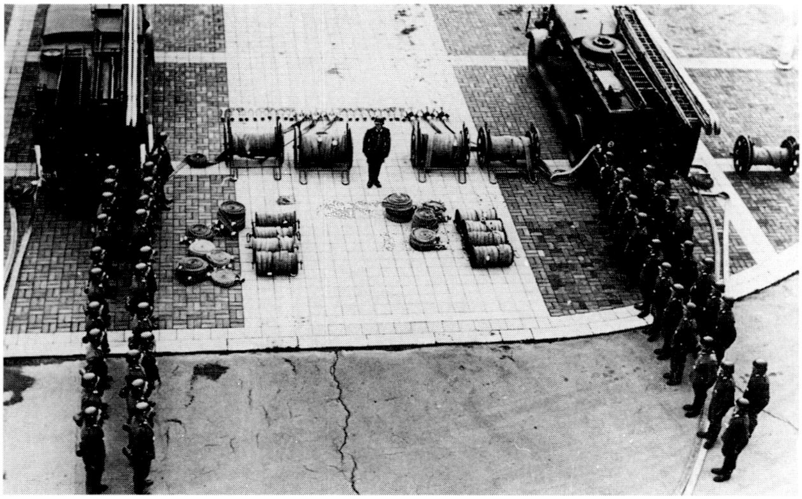

*Before the Olympic Stadium in Berlin, 1944: "Water lane" practice. The hose reels are unloaded and the crews lined up.* Photo: FMB.

*The firemen draw the hose lines forward. The first sprays of the "water lane" go to the left and right.* Photo: FMB.

*The water lane is completed.* Photo: FMB.

*Six LF 25 and eighteen firefighting groups create the water lane.* Photo: FMB.

*An LF 15 takes water from a hydrant, maintaining water supply for three firefighting groups.* Photo: FMB.

*The drill is finished, the hose rolled up and stowed in the LF 25.* Photo: FMB.

*LF 25 trucks supply water for the water lane. All the vehicles are equipped with cutterrs.* Photo: FMB.

*All the vehicles (LF 25, LF 15, LF 8 with TSA, tankers and cars) have turned out for a parade.* Photo: FMB.

*General Goldbach, 7th from left, inspects the Fire Police demonstration.* Photo: FMB.

*The firefighting crews stand before their trucks.* Photo: FMB.

The "Volunteer Firefighting Regiment" of East Prussia had been transferred to Warsaw at the end of 1941. When the unit was disbanded on June 14, 1943, the regiment was changed into FSchP Units 6 and 7. Later FSchP Unit 7 saw service in Berlin-Köpenick. In the summer of 1944, it was again ordered to transfer to Warsaw, where an uprising had meanwhile broken out. Two companies and their vehicles moved quickly to the Polish capital, but did not extinguish any fires there. Instead of that, the firemen were given flamethrowers and briefly trained to use them. Various units were assigned groups of eight firemen with flamethrowers, with which they set the city of a million inhabitants on fire. An army engineer remembers that firemen with flamethrowers had been assigned to his unit and burned prisoners and civilians alive.

The rebels had also set up a fire service, which even used vehicles, trucks with trailers. German vehicles were also captured and used.

# The Warsaw Rebellion of 1944

*Rebels in Warsaw (August 1944) remove a portable fire pump from a captured three-ton Opel LF 8 (with pressed-wood body).* Photo: Batuk.

In 1944 a plan to use firefighting tanks against large-scale fires and fires in impassible rubble areas was considered. Various captured vehicles were rebuilt for this purpose: the Italian Fiat Ansaldo L 33/35 tank; the Soviet T-20, and the German Sd. Kfz. 251/16 halftrack. These vehicles were to be displayed to a group of experts at Luckenwalde on June 8, 1944. The Sd. Kfz. 251/16 was to be equipped with a 700-liter water tank, a portable pump and two swiveling nozzles. According to Brunswig, it was intended that this vehicle be "a kind of water lane" to rescue people from large fires. But it was not destined to see service.

On the other hand, plans for flamethrowing tanks could be realized, including use of the Sd. Kfz. 251/16. On August 22, 1941 the OKH wrote to Koebe and shared available quantities of materials to rebuild ten armored vehicles (5828 kg). Units of the Army and the Waffen-SS directly ordered weapons or replacement parts for the Sd. Kfz. 251/16 flamethrowing armored vehicle or sent experience reports to Koebe. The special branch for flamethrowers and the Army Weapons Office urged Koebe to consider moving their production facilities in case of damage. Kumagnamed the Wumag A.-G. of Görlitz as an alternative location, since that firm "produces the vehicles and sends the same to Luckenwalde for installation of the systems."

On June 6, 1944 a Waffen-SS unit reported "that the flamethrowing tank Sd. Kfz. 251/16 had several easily vulnerable spots." The mantles of the steel barrels were penetrated by shell hits, "even the smallest calibers, near the vehicle's body, as well as by rifle and machine-gun fire."

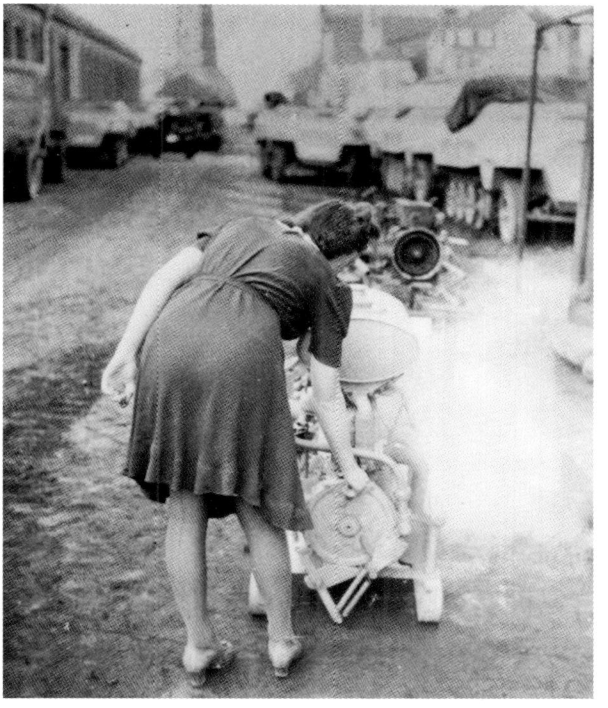

*At the Koebe firm's plant in 1944, a portable pump, planned by Koebe to be easy to use (even by women), is shown, with the Sd. Kfz. 251/16, which was to be fitted as a flamethrowing tank, in the right background.* Photo: Nase Collection.

# Flamethrowing and Firefighting Tanks

The last flamethrowing tanks, which had reached the division in March 1944, had shown numerous material damage. Tombac tubes, for example, had burst at only 13 atmospheres of pressure. Mauser barrels were ordered again and again.

*Design for a firefighting tank for a three-man crew, with cutters, 1944.* Photo: Nase

*A KS 25 of the Bremen Air-Raid Regiment (Magirus long-nose) is ready to go into action after the 111th air raid on Bremen, April 1943.* Photo: StaA Bremen.

*After an air raid on central Berlin in 1945, an LF 25 is seen in the background, with a TS and TSA of the SHD to the right.* Photo: SLBSt Berlin.

*March 13, 1945: An LF 25 stands in Wilhelm Square in front of the destroyed Reich Propaganda Ministry.* Photo: BAK J 31336.

*After an air raid on February 3, 1945, the people of Berlin are supplied with water with the help of fire trucks.*

*After an air raid on the VW city of Fallersleben/Wolfsburg, probably in the summer of 1944, an LF 25 (with wolf emblem) stands between VW Kübelwagen that were moved to safety.*
Photo: Fritz Heidrich/Wolfsburg City Museum.

*After the serious air raid on the Magirus-KHD works in Ulm on August 9, 1944, DL 26 trucks that were being built standing in lines were destroyed.*
Photo: Magirus-Iveco Archives.

# Destruction of Fire Apparatus and Truck Production

The Allied air attacks in the latter half of 1944 were directed at the motor vehicle industry and its related supplier industries. From July to December 1944, this branch of manufacturing was almost completely destroyed in the entire Reich area. The manufacturers of firefighting apparatus also had to endure heavy damage, extending to the destruction of their production facilities. Six of the ten producers of heavy and medium truck chassis were struck with 6530 tons of bombs in twelve pinpoint bombing attacks. Eighteen other air raids, with 3592 tons of bombs dropped, were directed at a further six motor vehicle manufacturers. The three greatest truck manufacturers (Opel, Ford and Mercedes) still produced some 4500 vehicles per month as of July 1941, but this production fell back drastically in November 1944.

On October 4, 1943 the facilities of Henschel in Kassel had already suffered severe damage from an air raid by 501 bombers. The Büssing-NAG factories in Braunschweig were also attacked intensely: on February 10 and April 8, 1944 the 8th USAAF attacked Braunschweig and caused severe damage to the two factory complexes. Also on April 8, 1944, 59 bombers of the 8th USAAF attacked the VW manufacturing plant in Wolfsburg in a daylight raid. The attacks of the 8th USAAF on the VW works were repeated on June 20, June 29 and August 5, 1944, with a total of 263 four-engine bombers. On August 5 and October 22, 1944, the Americans carried out further daylight raids, doing severe damage to the Büssing-NAG works on Wolfenbüttler Street, which had to stop production after the October 22 raid.

On August 6, 1944, the Opel truck factory in Brandenburg/Havel was destroyed by an American daylight raid; on August 13 the main plant in Rüsselsheim was struck. 50% of the building and 20% of the production capacity in Brandenburg and 50% of the plant in Rüsselsheim were destroyed.

The production facilities of the Daimler-Benz AG in Gaggenau were largely destroyed by air raids on September 10 and October 3, 1944.

The Metz firm in Karlsruhe was seized for war production by the Wehrmacht (Army Group G) in the middle of 1944, and on December 4 the factory was practically destroyed. In 24 minutes, 2297 tons of bombs, including 185,000 incendiary bombs, were dropped on the city and the adjoining industrial facilities. The factory firefighters could not extinguish the fires, and Factory I was 80% destroyed. Soon shortages of energy, raw-material supplying and transport facilities made themselves felt more and more clearly. At the end of 1944 and in 1945, the destroyed factory premises were often attacked by low-flying Allied planes, which caused additional damage.

A similar fate befell Magirus in Ulm. This firm had also moved some of its production out of Ulm since 1942 on account of the increasing danger of air raids; some of its fire truck production was moved to Buchau. In a daylight raid by 100 American bombers on September 13, 1944, the Söflingen factory was destroyed and had to give up production.

Factory I, 36,362 square meters in extent, lay in the center of Ulm and was first damaged slightly by an air raid on March 15, 1944. The attack on July 19, 1944 also did little damage to the works. In the air raid of 243 four-engine RAF bombers on December 17, 1944, several buildings were severely damaged by roof fires. The factory was extensively damaged on February 25, 1945, and after the air raid several buildings had to be written off as a total loss.

Factories II and III lay at the edge of the city and included an area of 78,361 square meters. The attack by Allied bombers on August 9, 1944 was directed specifically at these Magisur factories II and III, resulting in 17 dead, 40 injured and the total destruction of numerous buildings. After the attack on September 13, 1944 production was more or less stopped and the work force removed.

On December 17, 1944 the spare-part storehouse and three other buildings were completely destroyed.

On March 4, 1945, 500 American bombers attacked the wood-supply warehouse, and the entire stock of wood stored there was lost.

At the end of 1944, there was practically no facility left for the roduction of fire-truck chassis. The production facilities of Opel, Mercedes, Büssing-NAG and Henschel lay in rubble and ashes. The production of some 9100 trucks of all classes per month in the first half of 1944 sank, after the severe air raids, to 3202 in December 1944, mainly trucks for the Wehrmacht, including the "Mules" built by Magirus. Raw-material shortages and the destruction of traffic routes by Allied air raids brought about the end of fire-truck production at the end of 1944. The Chief of the German Police reported on January 21, 1945 that the production of fire trucks had been halted and communities should contribute fire trucks to regions threatened by air attack.

*The destroyed factory facilities of the Adam-Opel-AG in Rüsselsheim after the air raid in July and the RAF raid on August 13, 1944, in which 967 tons of bombs were dropped. Workers are covering the damaged roofs temporarily with tarpaulins.* Photo: BPK.

*The destroyed Mercedes-Benz truck factory in Gaggenau.* Photo: MB Archives.

*An aerial ladder truck destroyed in the final combat in Berlin, 1945.* Photo: FMB.

*Wilhelmshaven 1945: Two destroyed LF25 of the Luftwaffe.* Photo: BF Archives, Wilhelmshaven.

**Hamburg Fire Police, May 5, 1945**

"On May 5, 1945 at 4:35 PM, there appeared in the garage of Fire Station 12 four foreign civilian workers, presumably Poles or Ukrainians, and began to work on Motorcycle Pol 129332. The foreigners were ordered out of the building by Fire Captain Schmidt. At this time they resorted to force, struck the Captain in the face and left with the motorcycle.

About 5:45 PM four more foreign civilian workers, again Poles or Ukrainians, came and wanted to take two cars that were in service for the fire department, out of the building. Several members of the fire department tried to chase the foreigners out of the firehouse, whereupon the foreigners used force and struck our men with their fists. At the same time, British street sentries were called upon for help and informed that the vehicles belonged to the fire department for its work, whereupon the British soldiers intervened energetically, brought the foreigners out of the firehouse and sent them away."

**Reichsführer SS, June 10, 1943**

"The continuing production of LF 8, LF 15 and LF 25, in view of the air situation, is directed almost exclusively to the following localities: [a through f]

a) Luftwaffe for LS Units (mot.) airfield firefighters and Fire Police units;

b) Communities in regions that are exposed to constant air raids;

c) Factory air-raid units of firms with war-essential production, even when they are not in particularly air-threatened regions;

d) Communities in the area of which industry and production vital to the war exist;

e) BdO in the East and Ukraine, but to a limited extent;

f) Occasionally also in foreign countries, when particular conditions exist."

All objections would be pointless. For ladder trucks, hose trucks and trailers there would be no more apportioning, "since production is meager and already spoken for" (signed Goldbach to the HSSPF).

**The Local Air Raid Leader in Nürnberg, November 1944**

". . . About the use of motorized firefighting forces by the Luftwaffe and Fire Police, partly incorrect conceptions presently exist among the people. It is often claimed that many fire vehicles have just stood around during the whole period of action, without going into service. The investigation in such cases has almost always shown that (a) supply vehicles (repair-shop trucks, tank trucks, baggage trucks, etc.) had been confused with fire engines (such supplies must be taken along by the motorized firefighting units, because these units often do not return to their bases for days or weeks at a time), or b) the trucks were fire engines whose pumps were not needed for supplying water . . . Of a motorized air-raid unit, which possesses 36 firefighting vehicles, 16-18 fire engines are usually not needed to pump water as long as there is plenty of water available. Therefore the people who do not know the situation often get the impression that not all the available means are being utilized."

**F. Unkel about the action of the Volunteer Firemen of Marburg 1943-1945**

"On these trips we received horrifying experiences of the power of destruction. On account of our time-consuming approach routes of two or more hours, the worst was already over on our arrival, many houses were burned out or caved in, and sometimes there were pockets of fire that we had to drive through in order to reach the object to which we were directed. So it happened that we, held up by delays in the streets, had to stop beside such a burning house and then were surrounded by people and beseeched to help them and at least save the neighboring house. It was terrible to have to make clear to these people that our hands were tied. Thus it then occurred that the object to which we had been sent had meanwhile burned down and there was nothing more to extinguish there . . ."

# The End

After the Red Army had crossed the border into East Prussia, the flight from the East and the lost of the greatest part of the firefighting vehicles stationed there began. On January 23, 1945 the Commander of the Fire Police in Elbing, Major Isermann, gave the guiding order that the fire vehicles belonging to the Luftwaffe had to leave the city. This order was carried out. But with these vehicles, there also left, without order, all the city's own firefighting vehicles.

This withdrawal was serious to the extent that more than 25,000 people remained in the city and the events of the war caused numerous fires. On the evening of January 23, 1945 the first Russian tanks rolled into Elbing.

The same thing happened in the West: as the Americans reached the Rhine and threatened Cologne, all the fire department vehicles were moved to the east side of the Rhine, and Cologne was left without any fire protection!

The Russians moved forward toward Danzig, Stettin, Glogau and Breslau. Numerous firefighting vehicles could be rescued from Danzig and Stettin; some of them gathered at the premises of the main fire station in Hannover. In encircled Glogau, despairing firefighters had tried to extinguish fires with red wine for lack of water. Breslau was encircled and declared besieged; numerous vehicles belonging to the fire police remained in the city. After the war, the Soviets turned over to Dresden a number of Breslau vehicles. Because of the attacks on February 13 and 14, 1945, Breslau had had to accept particularly severe losses of firefighters: the loss of the entire Fire Police and the FE Service with 171 dead (including 129 Ukrainians), 300 missing (including 99 Air-Raid Police members from Silesia), and 132 wounded. Entire firefighting groups were wiped out, and on the evening of February 13, 1945 alone, 70 firefighting vehicles were burned in the firehouse by the Grand Park.

As the fronts moved toward the Oder and Rhein, the firefighters were used in direct combat actions. Some 20 percent of the firefighters had to work in the equipment industry outside of air-raid alarms. In January 1945, 250 Berlin firefighters were turned over to a police regiment and transferred to the eastern front. Hose trucks were turned over to the Wehrmacht in order to mobilize the anti-aircraft batteries of Berlin. On March 14, 1945, 125 men out of each group area, about one company, had to be turned over to form a flamethrower battalion. The flamethrowers were turned over by engineers, who also undertood training with the device. The unit was divided among the individual units of the city's defense, and most of them fell in the final combat around Berlin.

After the Soviets had crossed the Oder and advanced toward Berlin, members of the FE Service built barricades and supplied the population with water. Fire trucks towed 8.8 cm anti-aircraft guns to the East and brought ammunition to the front lines.

Rumors spread in Berlin: In Küstrin the German fire trucks had been destroyed by the Soviets. A supply of fuel was assembled secretly in order to let the valuable Berlin fire trucks and their crews escape to the West. The Berlin Commander of the Fire Police, Goldbach, had arranged the removal of the fire trucks. "In a discussion with all Fire Police group commanders" it was written in "125 Years of Berlin Firefighting", "Goldbach made it clear at the main fire station on April 22, 1945 that the remainders of the Groups North and South should first withdraw to Nauen, where they would be given further written marching orders. The two groups thereupon drove off to Nauen on the same day with a portion of their vehicles, which more or less corresponded to their peacetime strength. Vehicles of other units joined in. In Nauen, the Group East with some fifty vehicles received marching orders to go to Kellinghausen and Itzehoe. Via Schwerin (where wounded men from hospitals being transferred to the West were taken on) and Lauenburg, despite fire from low-flying planes, their goals were reached. The Group North with some forty vehicles received marching orders via Lübeck to Pinneberg. They also reached their goal. Both groups waited there for the arrival of the British."

Hundreds of firefighters from factory and Wehrmacht firefighting units had taken this route through Nauen. Supply Minister Albert Speer, who had been on the way to threatened Berlin, was amazed to see the masses of fire vehicles coming toward him. In the anniversary publication "125 Years of Berlin Firefighting" it also states: "Other units of the FE Service likewise wanted to move westward and began to move out without orders. On the Frey Bridge one unit of the Schöneberg Fire Department encountered a barricade that had been erected by the Wehrmacht, from which the vehicles with blue lights driving by were fired upon. The fire vehicles then had to turn back with their wounded members. Their attempt to take their wounded men to hospitals was in vain, since guards were posted before all the hospitals and, because of suspected shirking of duty, would not let anyone else in."

Numerous firefighters set out, wanting to avoid the impending encirclement of Berlin by the Soviet armies. SS courts-martial and military police searched through the ruins of the capital city, and hanged anyone suspected of being a deserter on the nearest lamppost. On April 26, 1945 there appeared in Goldbach's office at the Fire Headquarters on Linden Street a police command that wanted to take the General prisoner. When Goldbach reached into his jacket pocket to get a cigarette, one policeman shot him; the bullet hit Goldbach in the eye. Badly wounded, he was taken to the Police Hospital on Scharnhorst Street and executed there.

Of the last hours of the Berlin Fire Department, it is written in the 125-year anniversary publication: "Meanwhile the conquest of Berlin by the Russians had begun. The members of the Schöneberg brigade sat in the cellar as the Russians occupied their building and its yard. One of the men went into the courtyard with a white flag. Thereupon the Russians hauled all the remaining men out and took them prisoner. At the Lietzow station, three men left the cellar to surrender to the Russians. They encountered drunken Russians who shot two of them; the third was badly injured but survived . . . Gradually the front moved forward past the firehouses, whereby many came under infantry and artillery fire and were destroyed or damaged. A number of men and women helpers lost their lives in the process, and many vehicles were destroyed. The main fire station was occupied by the Russians on April 30, 1945. When Berlin surrendered on May 2, 1945, the last members of the FE Service were taken prisoner . . ."

*Dresden after the air raids on February 13 and 14, 1945: a ladder truck, on its way to a fire on the Feldgasse, was hit and burned out.* *Photo: Saxon State Library.*

**Firefighting Vehicles as of January 15, 1945:**

| Vehicle | National (gray) | City (green) | total |
|---|---|---|---|
| LF 25 | 146 | 56 | 202 |
| LF 15 | 75 | 49 | 124 |
| LF 15 (P.E.) | 8 | 0 | 8 |
| LF 8 | 138 | 20 | 158 |
| LF 8 (P.E.) | 22 | 0 | 22 |
| TLF 15 | 18 | 0 | 18 |
| S 4, 5 & 3 | 59 | 7 | 66 |
| Hose Trucks | 12 | 0 | 12 |
| DL 26 | 4 | 45 | 49 |
| DL 46 | 2 | 0 | 2 |
| Equipment Trucks | 0 | 16 | 16 |
| E-LKW | 143 | 0 | 143 |
| Generator Trucks | 2 | 0 | 2 |
| Special Body Types | 1 | 0 | 1 |
| Chiefs' Cars | 0 | 1 | 1 |
| Rescue Vehicles | 0 | 6 | 6 |
| Anti-Gas Trucks | 0 | 1 | 1 |
| RKW | 0 | 3 | 3 |
| PKW | 113 | 67 | 180 |
| Motorcycles | 391 | 7 | 398 |
| LKW | 0 | 20 | 20 |
| Buses | 0 | 1 | 1 |
| Trailers | 168 | 45 | 213 |

In the final combat, firefighters from the Ranke Street firehouse near the Charlottenburg Castle saw service; many were lost when they tried to cross the Courthouse Square on their return. The Hafenplatz firehouse was fought over for four days and occupied by the Russians on April 30, 1945. Practically all of the firemen of this station were lost, and the building was completely destroyed. At the Stettin station, an LF 25 and a DL were found after the capitulation. The fire service was at first resumed with two electric vehicles, in order to circumvent the lack of motor fuel. In July 1945, there were no more than ten fire trucks in Berlin!

| Vehicle | West Berlin | East Berlin | on 11/21/1948 |
|---|---|---|---|
| LF 25 | 29 | 9 | |
| LF 15 | 7 | 18 | |
| Equipment trucks | 15 | 16 | |
| Ladder trucks | 15 | 6 | |
| Hose trucks | 10 | 0 | |
| LKW | 12 | 7 | |
| Slw | 2 | 3 | |
| PKW | 16 | 7 | |
| Other | 13 | 5 | |
| Flb | 4 | 2 | |
| Total | 123 | 73 | |

In Oldenburg, the British military government took over two fire trucks, two pump trailers and three cars in June 1946. One of the trucks was found near Oldenburg but had to be junked because of irreparable damage. Four other trucks had already been more or less seriously damaged in the war. Two could be repaired, but what with the lack of spare parts, one of the other two trucks had to be junked. The Oldenburg City district had one LF 25, six LF 15, eight LF 8, one TFL 8, ten TS 8 and five TSK, plus two ladder trucks (DL 26 and DL 20), one equipment truck, two cars, one truck, and one service truck for electric-line repairs. In Wilhelmshaven the German firefighters had the following fire apparatus in February 1946 (in parentheses: ready for service): one LF 25 (1), nine LF 15 (1), one LF 8 (0), three cars (1), 5 motorcycles (3) and two other vehicles (2).

At the war's end in Kassel there were two LF 25, one DL 26 and one 4.5-ton SK available, plus one TFL, one 3.5-ton SK and one LF 8 in reserve, needing repairs.

On April 12, 1945 Celle was occupied by the Allies. The personnel of the provincial firefighting school in Celle who could be located were arrested and sent in the direction of Wietzenbruch. Meanwhile, the motor pool was plundered by Polish displaced persons. On April 13, though, all the vanished fire trucks could be gathered again and placed under guard. After the Bergen-Belsen concentration camp was liberated, the firefighting forces of the Celle school were sent to Belsen to supply the camp with water.

In the last years of the war, the volunteer firemen of Celle had driven their fire trucks into bomb shelters during air-raid alarms. The members of the Celle volunteer firemen were also taken prisoner in April 1945, but soon all who had not been party members were released. Here too, fire vehicles were utilized, partly by the British. In a short time, the greater part of the firefighting fleet could be assembled and made ready for service. Even at the end of the fifties, a ladder truck that disappeared in 1945 was brought back from Britain and returned to Celle.

Firefighting forces from Alsace came out of the West and arrived in Stuttgart. Two teams of the FE Service (many of their members Ukrainians) were sent marching in the direction of Lake Constance. They took sixteen LF 25 and two 4.5-ton SW, but two weeks later, on April 16, 1945, they were brought back to Stuttgart. There had been no combat in Stuttgart, and the fire officials were not arrested, as those in Karlsruhe had been, nor were their fire trucks requisitioned.

At the end of November 1945, Stuttgart's fire department had 93 vehicles: three cars, six cycles, 19 LF 25, 15 LF 15, 15 LF 8, 9 TFL 15, four rescue/equipment vehicles, four 4.5-ton hose trucks, ten ladder trucks, six trucks, one water rescue truck and one other special vehicle. Later the DL 46 from Nürnberg reached Stuttgart.

In Nürnberg, on the other hand, members of Fire Police Unit 8 had fought against the Americans in the streets, losing numerous men.

The BdO for Defense Zone IX (Kassel) had ordered early in March 1945 that all fire vehicles be taken away, to keep them away from the enemy, which meant that Kassel and its inhabitants no longer had any fire protection as of April 1, 1945. "What with attacks from low-flying planes, machinery damage that could not be repaired, lack of fuel, lack of tires, etc., a great number of fire trucks and equipment had been left along the roads. What little remained had to be hidden hastily in the Bohemian Forest near the Czech border, while their crews were taken prisoner. The trucks were taken out of their hiding places by the occupying troops, driven around in the area, and sometimes left to their fate in ravines, quarries and woods, after most of their tires had been destroyed with axes. Now the fire trucks and equipment were left for the people to plunder. There were also plenty of truckers and other individuals who quickly appropriated the individual vehicles and hauled them away." In this process of evacuation, one truck after another was abandoned, while twenty remained where they belonged: in Roitsch was DL No. 3, in Waldsassen two LF 8. In the woods near Waldkirch, three government-owned cars, four motorcycles, 24 pumpers, three hose trucks, four TFL and three other trucks had been left behind.

Nevertheless, some vehicles that had been removed were returned to Kassel, including the city's DL No. 2, a KS 25 and two hose trucks (one 4.5- and one 3.5-ton).

The list of the fire trucks that had been lost in April 1945 shows 23 vehicles that were taken away and 20 that were total losses.

"To search for what had been left in the Bohemian Forest in April 1945, two search commands were sent out in short order, and they found six city and five government fire trucks. The others had been bought at very low prices by private truckers and quickly rebuilt into cargo trucks. The return of these vehicles was specifically required by all the applicable agencies of the military and civilian governments. Without the authorization of the military government in Weiden/Oberpfalz and Selb, Bavaria, though, nothing could be accomplished." In 1946 the Kassel Fire Department again had fifteen fire trucks, including two DL 27 and two ladder trucks dating from 1914. There were also, among others, six water-cannon units with 2000, 1500 and 1200 liters-per-minute capability. It was decided by the Allies who arrived in Kassel on April 2 and 3, 1945 that a fire department for the city of Kassel could be put together from the factory fire brigades of the Henschel and Fieseler works. Henschel contributed one LF 25 and TFL, fieseler one LF 25. In all, there were two LF 25 units, one DL 26 and one 4.5-ton hose truck available, plus two LF 25, one TFL and one hose truck in reserve. In addition there were one car and one telegraph service truck. By February 1, 1948 the Kassel Fire Department again had 35 vehicles, including one DL 26. One of the lost DL 26 trucks was found badly damaged and turned over to the Metz firm for repairs.

In Austria too, fire trucks had to be located and brought back after the capitulation. On August 12, 1945, 52 trucks of the Vienna Fire Deparmtent had been assembled. In Styria, the British military government had turned 120 trucks and 15 towing tractors over to the fire departments by October 1946.

The chronicles of the Salzburg State Fire Department state: "The supply of vehicles and fire apparatus was more than complete in the State of Salzburg, thanks to the forced disarmament of the German Wehrmacht, and included then, as now, about forty vehicles."

*This old Magirus personnel vehicle with a TSA 28 trailer went back into service in 1945. Note the painted-out "Feuerschutzpolizei" lettering and police emblem, as well as the lack of a license plate. The Allies had ordered in Berlin that the city's fire vehicles had to have yellow front fenders and motor hoods, as well as a yellow circle on the rear.* Photo: FMB.

*The veteran "Cat" (based on its old KzS 8 designation), here marked LF-8/ 37-248.* *Photo: FMB.*

*Service Truck No. 56 of the Spandau Fire Department after 1945. The yellow circle on the rear is easy to see.* *Photo FMB.*

*A service truck of the Berlin Fire Department with a wrecked truck in 1946.* *Photo: LAB.*

*A Magirus pumper of the Berlin Fire Department in 1946.* Photo: FMB.

*A DL-46 ladder truck.* Photo: FMB.

*A Metz DL 46 ladder truck (+ 2 m) on Mercedes chassis in 1946. Here Russian lettering (BG-02-19) has been added. Note the fire extinguisher on the right side of the body.* Photo: FMB.

***A training mission of the Schöneberg Fire Department in Berlin, with pumpers and ladder trucks.*** ***Photo: FMB.***

*A fire drill at the main firehouse on Linden Street, Berlin, on September 28, 1948. After three years of progress under Fire Chief Feierabend, the Berlin professional firefighters had 34 fire companies and 49 volunteer units. In October 1948 their vehicles included 207 pumpers and special vehicles, including two rescue trucks and two special equipment trucks with heavy engineering equipment.* Photo: BAK 19000/2713.

*The "yellow period" is over. The Berlin Fire Department celebrated its 100th anniversary on Levetzow Street in the Zoo district on June 28, 1951. In front is an LF 25, then a Büssing-chassis ladder truck, then another LF 25. Note also the license plates (KB for Kommandeur Berlin).* Photo: LBSt Berlin.

*1960: Vehicles of the Oberlahnstein Volunteer Firemen. Note the rebuilt US Jeep, then an LF 8 on Opel Blitz chassis, an LF 15 in the middle with a pressed cardboard cab (built 1944-45) and, at right, an LF 15 with metal cab.* Photo: Gras Collection.

*A Mercedes LF 8 with front pump.* Photo: Gras Collection.

*Vehicles of the Beuel Volunteers in Bonn: a Mercedes ambulance and a Magirus LF 25.* Photo: StA Bonn.

# ABBREVIATIONS

| | |
|---|---|
| Abt. | Department |
| AL | Portable ladder |
| Ausb. | Training |
| BAK | Federal Archives, Koblenz |
| BdO | Police Commander |
| BF | Professional Firefighters |
| BLHA | Brandenburg State Archives |
| BPK | Prussian Cultural Photo Archives |
| Btl. | Battalion |
| CSSD | Chief, Security Police and SD |
| DL | Turntable aerial ladder |
| DM | German Museum (Munich) |
| EA | Electric Generator Trailer |
| FE-Dienst | Firefighting and Detoxifying Service |
| FF | Volunteer Firefighters |
| FM | Fire Museum |
| FSchP | Fire Protection Police |
| F.u.B.-Trupp | Firefighting and Rescue Troop |
| GDL | Large aerial ladder |
| GLG | Large Firefighting Group Vehicle |
| GRK | Large Equipment Truck |
| GSK | Large Hose Truck |
| HA | Main office |
| HJ | Hitler Youth |
| HSSPF | High SS and Police Leader |
| KBSt | District photo agency |
| Kdr. | Commander |
| KHD | Klöckner-Humboldt-Deutz |
| KL | Ladder Truck |
| Kmp. | Company |
| KS | Water cannon |
| KzS | Water-cannon truck |
| LBSt | State photo agency |
| LDL | Light ladder truck |
| LF | Pumper truck |
| LFV | State firefighting organization |
| LGK | Light gas-protection truck |
| LLG | Light fire pump |
| LS | Air-raid protection |
| MAN | Maschinenfabrik Augsburg-Nürnberg |
| MB | Mercedes-Benz |
| mot. | Motorized |
| OB | Lord Mayor |
| Orpo | Order-keeping Police |
| Pol | Police |
| PP | Police Chief |
| PW | Engineering truck |
| RA | Equipment trailer |
| RAF | Royal Air Force |
| RdErl. | Decree, orders |
| RFSS | Reich Führer SS |
| Rgt. | Regiment |
| RKW | Service truck |
| RLM | Reich Air Ministry |
| RMdI | Reich Ministry of the Interior |
| S | Hose truck |
| SA | Hose trailer |
| SLG | Heavy firefighting group vehicle |
| SRK | Heavy service truck |
| SSK | Heavy hose truck |
| SSPF | SS and Police Leader |
| StA | City Archives |
| StaA | State Archives |
| StM | City Museum |
| TLF | Tanker-pumper truck |

## Photo Credits

AMF (15), BF Archives, Göttingen (4), BF Archives, Hannover (2), Archives, Kassel (2), BF Archives, Cologne (15), BF Archives, Wilhelmshaven (1), BF Archives, Wolfsburg (1), Baluk, Warsaw (2), Photo Archives, Prussian Culture (7), Federal Archives, Koblenz (36), German Museum, munich (1), Eberth (1), Fire Museum, Berlin-Tegel (55), Heimberg/LFV Niedersachsen (18), Institute for City History, Frankfurt.Main (3), Johanssen (9), State Archives, Berlin (1), State Photo Agency, Berlin (5), State Photo Agency, Hessen (3), Iveco-Magirus Archives (47), Mercedes-Benz AG (8), Metz Archives, Karlsruhe (24), Mirko Nowak (2), Saxon State Library (1), Böning Collection (6), Fietz Collection (22), Eben Collection (2), Gerlach Collection (2), Göllner Collection (2), Gras Collection (6), Helmerding Collection (8), Hische Collection (3), Joop Collection (17), Koebe Collection (11), Nase Collection (2), Paulitz Collection (34), Unkel Collection (1), State Photo Agency, Hamburg (5), State Archives, Bremen (12), State Archives, Hamburg (1), City Archives, Bochum (12), City Archives, Bonn (2), City Archives, Braunschweig (3), City Archives, Düsseldorf (4), City Archives, Halle (1), City Archives, Kassel (2), City Historical Museum, Leipzig (4), City Archives, Mannheim (1), City Archives, Mönchen-Gladbach (3), City Archives, Mühlheim/Ruhr (6), City Archives, Munich (6), City Archives, Nürnberg (5), City Archives, Oberhausen (3), City Archives, Saarbrücken (2), City Archives, Speyer (4), City Museum, Wolfsburg (1), South German Publishing (2), Ullstein Photo Service (6), VW Factory Archives(1).

## Information and Interviews

Information: Zur, Ebel, Dietz, Heimberg, Rabe, Mack, Gerlach, Budzinski, Nase, Kreter.
Interviews: Wilhelm Bönig, 3/12/1994; Ewald Krone, 3/5/1994; Rudolf Joop, 10.20.1993; Hermann Günther, 3/11/1994; Albrecht Helmerding, 10/16/1993; Written records, Hans Hische, 1994.

## Archives

BF Kassel, State Firefighters' Organization of Niedersachsen, Nase, Bavarian State Archives, Munich, Brandenburg State Archives, Federal Archives, Koblenz, German Fire Museum, Fulda, Fire Archives, Berlin-Tegel, Magirus Firm Archives, State Archives, Berlin, State Fire School of Niedersachsen, Celle, State Archives of Niedersachsen, Hannover, Records of World War II Firefighting, Joop, Dietz Collection, Hische Collection, Unkel Collection, Bremen City Archives, Hamburg City Archives, Bonn City Archives, Braunschweig City Archives, Hannover City Archives, Wolfsburg City Museum, Metz Works Archives.

## Printed Sources (since 1945)

Brunswig, Hans, "Feuersturm über Hamburg", Stuttgart 1978; Dettmer, Werner, "Die Zerstörung Kassels im Oktober 1943", 1983; Dieckert, Grossmann, "Der Kampf um Ostpreussen", Munich, no date; Engelsing, Thomas, "Im Verein mit dem Feuer", Konstanz 1990; Giller, Mierke, Petig, :100 Jahre Berufsfeuerwehr Kassel 1891-1991", Kassel 1991; MacIsaac, David, "The United States Strategic Bombing Survey", Vol. 2, New York/London 1976; Metzger, Rolf, "150 Jahre Metz Feuerwehrgeräte", Bühl 1992; Obst, Dieter, "Reichskristallnacht" in Europäische Hochschulhefte, Frankfurt/Main/NY/Bern/Paris; Odziemkowski, Janusz, "Warszawa w wojnie obronnej 1939 roku", Warsaw 1989; Perschel, Rudolf, "Der Rote hahn über Braunschweig", Braunschweig 1955; Rabe, Klaus, "Der Zukunft ein Stück voraus: 125 Jahre Magirus", Düsseldorf 1989; Schläfer, Heinrich, "Die Münchner Feuerwehr", Munich 1979; Schmid, Armin, "Frankfurt im Feuersturm", Frankfurt/Main 1984; Schneider, Hans-Jürgen, "125 Jahre Opel Autos und Technik" 1987; Schramm, Georg Wolfgang, "Der zivile Luftschutz in Nürnberg 1933-1945" in Schriftenreihe des Stadtarchivs Nürnberg 1983, Vol. 35/I & II; Speer, Albert, "Erinnerungen", Frankfurt/Berlin/Vienna 1969; Strumpf, Günther, "Die Berliner Feuerwehr", Hanau 1987.
"Die Vertreibung der deutschen Bevölkerung aus den Gebieten östlich der Oder-Neisse, Vol. I in Dokumentation der Vertreibung der Deutschen aus Ost-Mitteleuropa, Munich 1984.
"Magirus Feuerlöschtechnik", Ulm 1953.
"Das Daimler-Benz Buch. Ein Rüstungskonzern im Tausendjährigen Reich", Nördlingen 1987, published by the Hamburger Stiftung für Sozialgeschichte des 20, Jahrhunderts.

## Printed Sources (before 1945)

Brunswig, Hans, "Das leichte Löschgruppenfahrzeug" in "Feuerschutztechnik" No. 4/1941; Garski, "Typisierte Feuerwehrfahrzeuge" in "Deutscher Feuerschutz" NO. 1, 1/8/1942; Garski, "Das schwere Löschgruppenfahrzeug" un Feuerschutztechnik No.9/41; Garski, "Der schwere Shclauchkraftwagen" in Feuerschutztechnik No. 10/41; Kalass, W., "Warschau" in Deutscher Feuerschutz No. 21, 11/1/1939; Kerstiens, Dr. Chr., "Das Reichsgesetz über das Feuerlöschwesen", Berlin 1940; Raue, "Der Stadtbrand von Kiev" in Deutscher Feuerschutz No. 21, 11/6/1941; Rumpf, "Das Pariser Feuerwehr-Regiment" in Deutscher Feuerschutz No. 18 of 9/16/1940; Schlosser, Karl, "Die Geräte der Feuerwehr", Munich 1943; Steiner, "Einsatzbereitschaft von Feuerwehrfahrzeugen und Maschinen im strengen Winter" in Feuerschutztechnik No. 7/1941;
"Anordnungen über den Bau von Feuerwehrfahrzeugen" (Vol. 1-4), Berlin 1940;
"Ausbildungsvorschriften für LS-Einheiten", Part I L. Div. 752/1a November 1943;
"Der Feuerschutz in Obenrost" in Deutscher Feuerschutz No. 21 of 11/1/1939.
"Kriegstagebuch der Feuerlösch- und Entgiftungsabteilung Tempelhof" (FMB sign. A 168a);
"Taschenbuch für Deutsche Feuerwehringenieure 1944", Berlin 1940;

Marburg, Marburg 1978.